Modeling and Applications in Operations Research

The text envisages novel optimization methods that significantly impact real-life problems, starting from inventory control to economic decision-making. It discusses topics such as inventory control, queueing models, timetable scheduling, fuzzy optimization, and the Knapsack problem. The book's content encompass the following key aspects:

- Presents a new model based on an unreliable server, wherein the convergence analysis is done using nature-inspired algorithms.
- Discusses the optimization techniques used in transportation problems, timetable problems, and optimal/dynamic pricing in inventory control.
- Highlights single and multi-objective optimization problems using pentagonal fuzzy numbers.
- Illustrates profit maximization inventory model for non-instantaneous deteriorating items with imprecise costs.
- Showcases nature-inspired algorithms such as particle swarm optimization, genetic algorithm, bat algorithm, and cuckoo search algorithm.

The text covers multi-disciplinary real-time problems such as fuzzy optimization of transportation problems, inventory control with dynamic pricing, timetable problem with ant colony optimization, knapsack problem, queueing modeling using the nature-inspired algorithm, and multi-objective fuzzy linear programming. It showcases a comparative analysis for studying various combinations of system design parameters and default cost elements. It will serve as an ideal reference text for graduate students and academic researchers in the fields of industrial engineering, manufacturing engineering, production engineering, mechanical engineering, and mathematics.

Jyotiranjan Nayak is a Professor in the Department of Mathematics at the Faculty of Science and Technology (IcfaiTech), IFHE, Hyderabad, India

Shreekant Varshney is Assistant Professor in the Department of Mathematics, School of Technology (SoT), Pandit Deendayal Energy University, Gandhinagar, Gujarat, India

Chandra Shekhar, the Professor and Ex-HoD of the Department of Mathematics at BITS Pilani, India

MATHEMATICAL ENGINEERING, MANUFACTURING, AND MANAGEMENT SCIENCES

Series Editor: Mangey Ram,

Professor, Assistant Dean (International Affairs), Department of Mathematics, Graphic Era University, Dehradun, India

The aim of this new book series is to publish the research studies and articles that bring up the latest development and research applied to mathematics and its applications in the manufacturing and management sciences areas. Mathematical tool and techniques are the strength of engineering sciences. They form the common foundation of all novel disciplines as engineering evolves and develops. The series will include a comprehensive range of applied mathematics and its application in engineering areas such as optimization techniques, mathematical modelling and simulation, stochastic processes and systems engineering, safety-critical system performance, system safety, system security, high assurance software architecture and design, mathematical modelling in environmental safety sciences, finite element methods, differential equations, reliability engineering, etc.

Biodegradable Composites for Packaging Applications
Edited by Arbind Prasad, Ashwani Kumar and Kishor Kumar Gajrani

Computing and Stimulation for Engineers
Edited by Ziya Uddin, Mukesh Kumar Awasthi, Rishi Asthana and Mangey Ram

Advanced Manufacturing Processes
Edited by Yashvir Singh, Nishant K. Singh and Mangey Ram

Additive Manufacturing
Advanced Materials and Design Techniques
Pulak M. Pandey, Nishant K. Singh and Yashvir Singh

Advances in Mathematical and Computational Modeling of Engineering Systems
Mukesh Kumar Awasthi, Maitri Verma and Mangey Ram

Biowaste and Biomass in Biofuel Applications
Edited by Yashvir Singh, Vladimir Strezov, and Prateek Negi

Modeling and Applications in Operations Research
Edited by Jyotiranjan Nayak, Shreekant Varshney, and Chandra Shekhar

For more information about this series, please visit: www.routledge.com/ Mathematical-Engineering-Manufacturing-and-Management-Sciences/ book-series/CRCMEMMS

Modeling and Applications in Operations Research

Edited By
Jyotiranjan Nayak, Shreekant Varshney, and
Chandra Shekhar

CRC Press
Taylor & Francis Group
Boca Raton London New York

CRC Press is an imprint of the
Taylor & Francis Group, an **informa** business

First edition published 2024
by CRC Press
2385 NW Executive Center Dr Suite 320, Boca Raton, FL 33431

and by CRC Press
4 Park Square, Milton Park, Abingdon, Oxon, OX14 4RN

CRC Press is an imprint of Taylor & Francis Group, LLC

© 2024 selection and editorial matter, Jyotiranjan Nayak, Shreekant Varshney, and Chandra Shekhar

ISBN: 978-1-032-40737-1 (hbk)
ISBN: 978-1-032-61187-7 (pbk)
ISBN: 978-1-003-46242-2 (ebk)

DOI: 10.1201/9781003462422

Typeset in Sabon
by Apex CoVantage, LLC

Contents

3 Different variants of unreliable server: an economic approach   61

SHREEKANT VARSHNEY, CHANDRA SHEKHAR, VIVEK TIWARI, AND
KOCHERLAKOTA SATYA PRITAM

13   **Food waste: impact of COVID-19 on urban and
     rural areas**   223

SURYA KANT PAL, MAHESH KUMAR JAYASWAL, SUBHODEEP MUKHERJEE,
KRISS GUNJAN, JYOTIRMAI SATAPATHY, AND SIMRAN SINGH

14   **A conceptual study of association factors contributing
     to the stress level of parents during the pandemic era
     in NCR**   235

NITENDRA KUMAR, POOJA SINGH, SANTOSH KUMAR, SURYA KANT PAL,
KHURSHEED ALAM, AND MAHESH KUMAR JAYASWAL

Acknowledgments

नमस्ते सदा वत्सले मातृभूमे
त्वया हिन्दुभूमे सुखं वर्धितोऽहम्।
महामङ्गले पुण्यभूमे त्वदर्थे
पतत्वेष कायो नमस्ते नमस्ते।।

This book is transcribed while the editors are supported by institutes of repute, IFHE, PDEU, and BITS Pilani. We acknowledge the heartiest thanks to these institutes' administrative and academic support. Very special thanks are offered to Prof. Mangey Ram, Series Editor; Mr. Gauravjeet Singh, Senior Commissioning Editor, CRC Press; Isha Ahuja, Editorial Assistant-Engineering, CRC Press, for their support and suggestions to improve the text. We are also indebted to the learned reviewers whose remarks were constructive and who thoroughly improved manuscripts. The contributors of a variety of manuscripts related to Operations Research and allied fields deserve a note of thanks for sharing content-rich thoughts. We are grateful to all well-wishers who enlightened the path of progress.

We thank our parents, children, friends, and society for inspiration, encouragement, and cooperation. Dr. Nayak is grateful to his wife, Ms. Namita, for her prompt support. Dr. Varshney dedicates this to his grandparents, Shri C. B. Varshney and Smt. Rammurti Devi. Prof. Shekhar acknowledges the support of his father, Shri L. P. Jayaswal, and wife, Ms. Shikha Gupta, for their understanding during time busy and continued motivation for the book's preparation.

At last, we are also thankful to all well-wishers who provided us with the best possible support, help, and backup in completing the work.

Jyotiranjan Nayak
Shreekant Varshney
Chandra Shekhar

Biography of editors

Jyotiranjan Nayak is a Professor in the Department of Mathematics at the Faculty of Science and Technology (IcfaiTech), IFHE, Hyderabad, India. Prior to this, he was Associate Professor at the Institute of Technical Education and Research, Siksha "O" Anusandhan University, Bhubaneswar, India, and has more than 32 years of teaching experience in undergraduate and postgraduate engineering studies. He taught the subjects like differential equations, vector analysis, complex analysis, linear algebra, numerical methods, optimization techniques, discrete mathematics, and advanced calculus at the undergraduate level. He taught postgraduate students algebra, cryptography, number theory, real analysis, topology, combinatorics, graph theory, and operations research. For Ph.D. students, he has delivered lectures on advanced computational mathematics, matrix algebra, and design of experiments at different universities, including Utkal University, Biju Patnaik University of Technology, Berhampur University, Siksha "O" Anusandhan, and ICFAI University. Besides, he has been a guest and visiting faculty at many other universities in the country. He obtained his Ph.D. degree from Utkal University in 2005. His research areas include non-convex programming, fuzzy optimization, optimal control, and topology optimization. He has published 12 research papers in leading national and international journals. He is also the former associate editor of *The Journal of Orissa Mathematical Society*. He was Vice-President of the Operational Research Society of India (ORSI), Bhubaneswar chapter. He is an executive member of the Orissa Mathematical Society (OMS) and a life member of the Indian Society for Technical Education (ISTE), Orissa Information Society (OITS), and Ramanujan Society of Mathematics and Mathematical Sciences (RSMSS).

Shreekant Varshney is Assistant Professor in the Department of Mathematics, School of Technology (SoT), Pandit Deendayal Energy University, Gandhinagar, Gujarat, India. Before this, he worked as Assistant Professor in the Department of Mathematics at the Faculty of Science and Technology (IcfaiTech), IFHE, Hyderabad, India. He obtained his

Ph.D. degree from the Birla Institute of Technology and Science Pilani, Pilani Campus, in 2020. He is actively involved in the research areas, namely queueing theory, machine repair problem (MRP), optimal control, reliability and maintainability, stochastic modeling, sensitivity analysis, evolutionary computation, statistical analysis, fuzzy set and logic, etc. He has taught undergraduate subjects such as linear algebra, differential equations, complex analysis, advanced calculus, numerical methods, and probability and statistics. He also has experience in teaching courses like research methodology, stochastic modeling, and introduction to queueing theory for Ph.D. students. Besides attending, presenting scientific papers, and delivering invited talks in FDPs, he has organized several workshops and symposiums. He has been awarded second prize in the academic writing competition organized by SIAM journal publishing. He has 13 research articles in journals of high repute, namely, *Reliability Engineering and System Safety, Journal of Computational and Applied Mathematics, Quality Technology and Quantitative Management, Arabian Journal of Science and Engineering*, etc. He is also a reviewer of many reputed journals. As a professional, he has visited IIRS (ISRO), CSIR-IIP, and WIHG.

Chandra Shekhar, the Professor and Ex-HoD of the Department of Mathematics at BITS Pilani, India, is actively involved in research and teaching in the area of queueing theory, computer and communication systems, machine repair problems, reliability and maintainability, stochastic processes, evolutionary computation, statistical analysis, and fuzzy set and logic. He has expertise in the subjects of probability and statistics, differential equations, linear algebra, advanced calculus, complex variables, and statistical inference at the undergraduate level of teaching. He has taught real analysis, topology, cryptography, applied mathematics, computational mathematics, cybernetics, and many more at the post-graduation level. He has been a pioneer in evolutionary computation, Markovian and stochastic modeling, research methodology, and queueing analysis and its applications. Besides attending, presenting scientific papers, and delivering invited talks at national/international conferences and FDPs, he has organized many conferences, workshops, and symposiums as a convener and as an organizing secretary. The best research paper award has been bestowed at the international conference. He has 48 research articles in these fields in journals of high repute and has supervised two Ph.D. theses. Besides some book chapters in an edited book published by the publisher of international reputation, authorship of the textbook titled *Differential Equations, Calculus of Variations and Special Functions* and the edited book titled *Mathematical Modeling and Computation of Real-Time Problems: An Interdisciplinary Approach* is also to his credit. He is also a member of the editorial board and a reviewer of many reputed journals, academic societies and Doctoral Research Committee, advisory board, faculty selection committee, and the examination board of many governments and private universities, institutions, or research labs. As a professional, he has visited IIRS (ISRO), CSIR-IIP, NIH, WIHG, CPWD, NTPC, Bank of Maharashtra, and APS Lifetech.

Contributors

M. M. Acharya
Department of Mathematics
School of Applied Sciences, KIIT
University
Bhubaneswar, India

Srikumar Acharya
Department of Mathematics
School of Applied Sciences, KIIT
University
Bhubaneswar, India

Altaf Ahmad
Faculty of Science and Technology
ICFAI Foundation for Higher
Education
Hyderabad, Telangana, India

Khursheed Alam
Department of Mathematics
School of Basic Sciences and
Research, Sharda University
Greater Noida, Uttar Pradesh, India

Rakesh P. Badoni
Department of Mathematics
Ecole Centrale School of
Engineering, Mahindra
University
Hyderabad, India

Sanjaya K. Behera
Department of Mathematics

Government Women's College
Sundargarh, Odisha, India

Vaibhav Bhartia
School of Computer Science and
Engineering, VIT
Vellore, Tamil Nadu, India

L. N. Das
Department of Applied
Mathematics
Delhi Technological University
Delhi, India

Jayanta Kumar Dash
Department of Mathematics, Siksha
"O" Anusandhan,
Deemed to be University
Bhubaneswar, Odisha, India

Subhakanta Dash
Department of Mathematics
Silicon Institute of Technology
Bhubaneswar, India

Kriss Gunjan
Department of Mathematics
Sharda School of Basic Sciences and
Research, Sharda University
Greater Noida, Uttar Pradesh, India

C. B. Gupta
The North Cap University

Gurugram, India
Mahesh Kumar Jayaswal
Department of Mathematics
Banasthali Vidyapith
Banasthali, Rajasthan, India

Topunuru Kaladhar
ICFAI Business School,
ICFAI Foundation for Higher
Education
Hyderabad, Telangana, India

Amit Kumar
Department of Mathematics
S.M.P.Gov. Girls. (PG) College
Madhavpuram
Meerut

Nitendra Kumar
Department of Decision Sciences
Amity Business School
Amity University
Noida, Uttar Pradesh, India

Sanjeev Kumar
Department of Computer Science
and Engineering
The ICFAI University
Dehradun, Uttarakhand, India

Santosh Kumar
Department of Mathematics
School of Basic Sciences and
Research, Sharda University
Greater Noida, Uttar Pradesh,
India

Vipin Kumar
Department of Mathematics
B. K. Birla Institute of Engineering
and Technology
Pilani, Rajasthan, India

Pradip Kundu
School of Computer Science and
Engineering
XIM University
Bhubaneswar, India

Mukesh Mann
Department of Computer Science
and Engineering
Indian Institute of Information
Technology
Sonepat, Haryana, India

Anjanna Matta
Department of Mathematics
ICFAI Foundation for Higher
Education
Hyderabad, India

Tusar Kanti Mishra
School of Computer Science and
Engineering, VIT
Vellore, Tamil Nadu, India

R. P. Mohanty
Siksha "O" Anusandhan University
Bhubaneswar, Odisha, India

S. P. Mohanty
Department of Mathematics
SOA University
Bhubaneswar, India

Ruchika Moharana
Department of Mathematics,
Siksha "O" Anusandhan,
Deemed to be University
Bhubaneswar, Odisha, India

Subhodeep Mukherjee
Department of Operations
GITAM Institute of Management

(Deemed to be University)
Visakhapatnam, Andhra Pradesh,
India

D. K. Nayak
Department of Mathematics
Veer Surendra Sai University
of Technology
Burla, Odisha, India

Jyotiranjan Nayak
Department of Mathematics,
Faculty of Science and
Technology
IFHE Deemed to be University
Hyderabad, India

S. K. Paikray
Department of Mathematics
Veer Surendra Sai University of
Technology
Burla, Odisha, India

Surya Kant Pal
Department of Mathematics
Sharda School of Basic Sciences and
Research, Sharda University
Greater Noida, Uttar Pradesh,
India

Kocherlakota Satya Pritam
Department of Mathematics
School of Technology, Pandit
Deendayal Energy University
Raysan, Gandhinagar, Gujarat,
India

Rajpal Rajbhar
Department of Applied
Mathematics
Delhi Technological University
Delhi, India

Manidatta Ray
Decision Science and Operations
Management,
Birla Global University
Bhubaneswar, Odisha, India

Anuradha Sahoo
Department of Mathematics, Siksha
"O" Anusandhan
Deemed to be University
Bhubaneswar, Odisha, India

A. K. Sahoo
Department of Mathematics
Veer Surendra Sai University of
Technology
Burla, Odisha, India

Manoranjan Sahoo
Department of Mathematics
School of Applied Sciences, KIIT
University
Bhubaneswar, India

Ashirbad Sarangi
Department of Computer Science
and Engineering
The ICFAI University
Dehradun, Uttarakhand, India

Jyotirmai Satapathy
Department of Mathematics
Sharda School of Basic Sciences and
Research, Sharda University
Greater Noida, Uttar Pradesh,
India

Chandra Shekhar
Department of Mathematics
Birla Institute of Technology and
Science Pilani, Pilani Campus
Pilani, Rajasthan, India

Pooja Singh
School of Business Studies
Sharda University
Greater Noida, Uttar Pradesh,
India

Simran Singh
Department of Mathematics
Sharda School of Basic Sciences and
Research, Sharda University
Greater Noida, Uttar Pradesh,
India

Admasu Tadesse
Department of Mathematics
School of Applied Sciences, KIIT
University
Bhubaneswar, India

Vivek Tiwari
Department of Mechanical
Engineering
Birla Institute of Technology and
Science Pilani, Pilani Campus
Pilani, Rajasthan, India

B. K. Tripathy
School of Information Technology
and Engineering, VIT
Vellore, Tamil Nadu, India

Shreekant Varshney
Department of Mathematics, School
of Technology
Pandit Deendayal Energy University
Raysan, Gandhinagar, Gujarat,
India

Ant colony optimization algorithm for the university course timetabling problem using events based on groupings of students

Rakesh P. Badoni, Sanjeev Kumar, Mukesh Mann, R. P. Mohanty, and Ashirbad Sarangi

1.1 INTRODUCTION

Timetabling is a significant and demanding research domain with a broad diversity of applications in several disciplines such as education, enterprises, transportation, human resources planning, sports, and logistics. Wren [1] proposed the standard definition of timetabling as the allocation, subject to constraints, of assigned resources to objects being positioned in space-time in such a way that a set of desirable objectives is satisfied as nearly as possible. These combinatorial optimization problems are multi-objective and high-dimensional and they receive exceptional attention from the research community as their manual solution is very time-consuming and requires innumerable effort [2]. Moreover, the obtained results are generally expensive in terms of resources and money, as they require a lot of human resources and computational power. In fact, a few minor changes also require starting the procedure over from scratch. It makes the automated timetabling requirement necessary, which reduces the time of formation, minimizes errors, and maximizes the satisfaction of desired objectives. One of the most extensively studied categories of the timetabling problem is the educational timetabling problem. A general and efficient solution to an educational timetabling problem is very complicated due to diversity, constraints variances, and changed requirements. The educational timetabling problem is broadly categorized into school and university timetabling problems. In turn, the university timetabling problem is divided into university examination and university course timetabling problems.

The university course timetabling problem (UCTP) is an NP-hard combinatorial optimization problem class where a set of events (lectures, tutorials, and laboratories) needs to be assigned into timeslots and suitable rooms according to students' enrolment data. The objective is to identify a consistent assignment that minimizes the violation of soft constraints and satisfies all of the hard constraints. The satisfaction of all the hard constraints

DOI: 10.1201/9781003462422-1

provides a feasible solution and must be satisfied under any situation. Although the violation of soft constraints does not influence the solution's viability, it is highly desirable to satisfy them to produce a higher-quality solution. As a result, soft constraint violations can be tolerated, but they are each subject to a penalty cost. Thus, the objective of the UCTP is to obtain a feasible timetable with a minimum overall penalty cost of soft constraint violations. If both the hard and soft constraints are satisfied, a solution of the UCTP is considered the optimal solution. The quality of a feasible solution is evaluated based on soft constraint violations. The lower the soft constraint violations, the better the solution quality is.

In this work, we propose a new idea of an ant colony optimization (ACO) algorithm using events based on students' groupings to solve the UCTP. First, students' mutually disjointed groups are formed based on selected events from a list of offered events. Here, each student belongs to exactly one group, and the task is accomplished by prohibiting the student from further selection once they are selected in a group. The union of all the events taken by the students of that group is assessed next. The group's cardinality is then evaluated, which refers to the group's total number of events corresponding to its group size. The cardinality of each group can be restricted by taking the maximum number of events per student as a lower bound and the total number of timeslots as an upper bound. This technique of forming disjoint student groups is revised until each group's cardinality does not lie within these bounds by not choosing common events among every group student. Instead of individual students, these students' group events are now assigned to timeslots and suitable rooms by employing an ACO algorithm. The proposed algorithm has experimented with several benchmark UCTP instances of various complexities to demonstrate its effectiveness in terms of the fitness function value. Every problem instance is run independently multiple times, and the solution with the lowest fitness function value among them is considered the solution of the problem instance. We observe that our approach performs competitively with those already available in the literature when compared in terms of the fitness function value. Because our approach uses events based on students' groupings with an ACO algorithm, we named it ant colony optimization with student groupings (ACOWSG). To the best of our knowledge, the proposed ACOWSG algorithm is the only existing algorithm available in the literature using the concept of grouping with an ACO algorithm.

The structure of this chapter is organized as follows. Section 1.1 covers the introduction. The background of the related work on the UCTP is briefly described in Section 1.2. Section 1.3 explains the considered UCTP along with its mathematical formulation. The ant colony optimization algorithm is discussed in Section 1.4. The combination of the proposed student grouping algorithm with the ant colony optimization algorithm for the solution of the considered UCTP is described in Section 1.5. In Section 1.6, the implementation and testing of the proposed algorithm for several benchmark problem instances of various complexities are carried out. Finally, conclusions are included in Section 1.7.

1.2 RELATED WORK

The history of the educational timetabling problem is older than five decades and started with [3]. Over the year, various researchers proposed several solution approaches and tested them on real-world problem instances. Although a significant development has been made in this domain, the research community has faced greater difficulty in comparing their algorithms with existing state-of-the-art algorithms. The main reasons were the different problem formulations and instances used by various researchers. To overcome this situation, the International Metaheuristic Network organized the First International Timetabling Competition (ITC2002)[1] in 2002. The objective was to simulate a real-world scenario where students have priorities to take the events according to their choices, and the timetable is constructed according to these preferences. Since then, these artificially generated course timetabling problem instances have become a standard within the research domain and have been used by several researchers to demonstrate their novel techniques' effectiveness.

During the last couple of decades, the UCTP has been widely studied, and significant research has been carried out by various researchers [4, 5, 6, 7, 8, 9]. These researchers proposed a large variety of exact and heuristic algorithms and applied them successfully to solve the problem. Initially, graph coloring algorithms [10, 11, 12] were widely used to solve these problems. In graph coloring algorithms, vertices and edges correspond to lectures and constraints, respectively, and then these algorithms are used for their solutions. In this kind of approach, a fixed number of colors equal to the number of timeslots are assigned to the nodes in such a manner that adjacent nodes must have different colors. These algorithms have demonstrated greater efficiency in small-sized problem instances, but their performance decreases progressively as the problem instance's size increases. Constraint-based approaches [13, 14, 15] have been extensively used for solving UCTPs. These approaches are defined as a set of variables (lectures, tutorials, and laboratories) to which values (rooms and timeslots) are assigned. The objective is to obtain consistent value assignments for variables to maximize the number of constraints satisfied. These constraint-based approaches are further hybridized with standard methods, such as integer, mixed-integer, and linear programming, to generate robust tools for their solutions. Several

---

1 http://sferics.idsia.ch/Files/ttcomp2002/oldindex.html

integer and linear programming approaches [5, 7, 16, 17] are also proposed for the solution of the UCTP.

The term metaheuristic, a category of heuristic techniques, was first introduced by [18]. Since then, they have become salient approaches among the other heuristic approaches for solving a wide variety of difficult combinatorial optimization problems. The motivation behind these metaheuristic approaches is nature-inspired evolution, and they are applied like evolutionary processes to find optimal or near-optimal solutions. Due to the extraordinary accomplishment in solving the UCTP, these metaheuristic approaches have been extensively published [19, 20, 21, 22, 23, 24, 25]. In general, these metaheuristic algorithms are classified as local area-based and population-based algorithms. Local area-based algorithms, also known as single-point algorithms, emphasize exploitation rather than exploration [26]. This implies that such types of algorithms find good solutions by making iterative changes to a single solution. However, such algorithms work in an unstructured way, which may lead to finding a solution in a single direction without checking an extensive search of the solution space until a stopping criterion is met [19]. Local area-based algorithms include Tabu Search (TS) [22, 24], Iterated Local Search (ILS) [9, 13], Simulated Annealing (SA) [14, 23], Great Deluge (GD) [27, 28], Variable Neighborhood Search (VNS) [29, 30], and many more algorithms.

On the other hand, population-based algorithms, also known as multiple-point algorithms, are good at exploration rather than exploitation [24]. These types of algorithms consist of several individual solutions that are maintained in a population. An appropriate selection procedure is further used at each iteration to select and update solutions in the population by creating new solutions and then updating them with the existing population. These algorithms refine the entire search space population to get a globally optimal solution and are sometimes referred to as global area-based algorithms. Such algorithms do not concentrate on the good fitness individuals within a population and search the entire solution space for the possible solution. Premature convergence is the main disadvantage of such types of algorithms. The more popularly used population-based algorithms used to solve timetabling problems include genetic algorithms (GAs) [31, 32], ant colony optimization (ACO) [33, 34], harmony search (HS) [35], artificial immune system (AIS) [36], artificial bee colony (ABC) [20, 21], and many more algorithms.

The ACO, proposed by [37], has been widely used in recent years by several researchers for the solution of the UCTP. Socha et al. [38] proposed a MAX-MIN ant system for their solution. The algorithm uses a different LS routine, which can be handled further by producing an appropriate construction graph. Here, the events' assignment to timeslots depended on the pheromone value within bounds. They further concluded that the MAX-MIN ant system performs significantly better than the random restart LS compared with a set of typical problem instances. The authors [39] studied the ACO algorithm along with four other metaheuristic algorithms. In their algorithm, each of m ants constructs a complete event–timeslot assignment by selecting event by event sequentially at each iteration. An event is further selected from a predefined list of events by an ant and scheduled to a timeslot in a probabilistic manner. The authors [40] developed a die-hard cooperative ant behavior approach (DHCABA) for the solution of the UCTP. They observed that DHCABA produced better results than the previous ant colony algorithms for most of the problem instances. The reason was the enforcement of a group of ants to walk along the desired path toward a feasible and good-quality solution. The authors [41] proposed two hybrid ant colony systems (ACSs) by hybridizing the ACS with SA (ACS-SA) and TS (ACS-TS) to solve the UCTP. In their algorithms, the ACS tries to find a feasible solution, and then the LS method using a large number of iterations is applied to minimize the violation of soft constraints. They concluded that ACS is robust and easy to combine with other methods such as SA and TS. The authors [34] used an ACO algorithm to solve the UCTP. The use of two distinct but simplified pheromone matrices is the essential feature of their algorithm to improve convergence. Furthermore, they suggested that the quality of the solution could be improved significantly by parallelizing the algorithm. The authors [25] proposed a hybrid approach by combining GA and ACO for the rescheduling problem in the middle of the running semester. Their approach's main feature was the minimization of errors by reducing clashes while substituting lecturers' rescues at mid-term. An ACO-based approach has been developed by the authors [42] for solving UCTP. They tested the algorithm over a set of data instances collected from various universities and obtained significant outcomes.

Reference [43] presented a grouping genetic algorithm for the feasible solution of the UCTP. They considered feasibility and optimality as two distinct sub-problems and proposed that any algorithm's performance for the fulfillment of soft and hard constraints might be different. In their approach, they used the concept of grouping proposed by [44] as the problem where the task is to partition an object set X into a collection of mutually disjoint subsets $x_i \in X$ such that $\cup x_i = X$ and $x_i \cap x_j = \phi$ for $i \neq j$, and according to a set of problem-specific constraints defining valid and legal groupings. Reference [45] presented a new hybrid algorithm using events based on the grouping of students for the solution of the UCTP. They also used the definition of grouping defined by [44]. Furthermore, they used the GA with LS to assign the events to timeslots and rooms. The resemblance between [43] and [45] is that they both used GA with their grouping approach.

In this chapter, we are trying to make an effort to utilize the concept of grouping along with an ACO algorithm for the solution of the UCTP. To the best of our knowledge, this approach is the only existing approach available in the literature using the concept of grouping with the ACO algorithm.

1.3 STATEMENT OF UNIVERSITY COURSE TIMETABLING PROBLEM

The UCTP models considered in this chapter were introduced in [38] and in the second track of ITC2007.[2] Here, we explain the UCTP along with its mathematical formulation. A UCTP instance's solution involves the assignment of a set E of events to 45 timeslots and a set R of rooms. These 45 timeslots are spanned over five days, where each day has nine timeslots. Furthermore, a set S of students attends these E events, and each room consists of a set of features from a given set F of rooming features. The objective is to obtain a feasible solution first by satisfying all the considered hard constraints and then minimizing the violation of soft constraints as far as possible to get optimal or near-optimal solutions. The formal demonstration of the problem is as follows:

Given

- $E = \{e_1, e_2, \ldots, e_n\}$ is a set of n events.
- $R = \{r_1, r_2, \ldots, r_m\}$ is a set of m rooms.
- $T = \{t_1, t_2, \ldots, t_{45}\}$ is a set of 45 timeslots.
- $S = \{s_1, s_2, \ldots, s_p\}$ is a set of p students.
- $F = \{f_1, f_2, \ldots, f_q\}$ is a set of q features of room.
- $CP(r_i)$ is the capacity of room r_i.
- A matrix $SE = \left[se_{ij} \right]_{p \times n}$, called student-event matrix and representing which event is attended by which student. Here, $se_{ij} = 1$, if student s_i is attending the event e_j; otherwise, the value is zero.
- A matrix $RF = \left[rf_{ij} \right]_{m \times q}$, called room-feature matrix and representing the feature possesses by the room. Here, $rf_{ij} = 1$, if room r_i is having feature f_j; otherwise, the value is zero.
- A matrix $EF = \left[ef_{ij} \right]_{n \times q}$, called event-feature matrix and representing the features required by the event. Here, $ef_{ij} = 1$, if feature f_j is required by the event e_i; otherwise, the value is zero.
- A matrix $ET = \left[et_{ij} \right]_{n \times 45}$, called event–timeslot matrix and representing the availability of timeslot for the event. Here, $et_{ij} = 1$, if event e_i can take place at timeslot t_j; otherwise, the value is zero.
- A matrix $EP = \left[ep_{ij} \right]_{n \times n}$, called event-preference matrix and representing the preference of one event over another event. Here, $ep_{ij} = 1$, if event e_i has to be scheduled before event e_j. Further it will take value -1 if event e_i has to be scheduled after event e_j. Finally, it will take value zero if there is no restriction.

---

2 www.cs.qub.ac.uk/itc2007/postenrolcourse/course_post_index.htm

- x_{ijkl} is a decision variable signifying student s_i taking an event e_j in timeslot t_k and room r_l and defined for $1 \le i \le p, 1 \le j \le n, 1 \le k \le 45$, and $1 \le l \le m$, as

$$x_{ijkl} = \begin{cases} 1 & \text{if the above combination holds} \\ 0 & \text{otherwise} \end{cases}$$

- y_{ijk} is a decision variable signifying an event e_i held in room r_j with feature f_k and defined for $1 \le i \le n, 1 \le j \le m$, and $1 \le k \le q$, as

$$y_{ijk} = \begin{cases} 1 & \text{if the above combination holds} \\ 0 & \text{otherwise} \end{cases}$$

- z_{ij} is a decision variable signifying an event e_i held in timeslot t_j and defined for $1 \le i \le n$, and $1 \le j \le 45$, as

$$z_{ij} = \begin{cases} 1 & \text{if the above combination holds} \\ 0 & \text{otherwise} \end{cases}$$

- a_{ij} is a decision variable signifying an event e_i scheduled before an event e_j and defined for $1 \le i \le n, 1 \le j \le n$, and $i \ne j$, as

$$a_{ij} = \begin{cases} 1 & \text{if the above combination holds} \\ 0 & \text{otherwise} \end{cases}$$

One of the objectives of the UCTP is to minimize the violations of the following three soft constraints. However, their violation is permitted, but at some penalty cost. In the following mathematical formulation, scv denotes the number of soft constraint violations.

Minimize:

$$\sum_{i=1}^{3} \text{scv}_i$$

1. Penalty for students with an event scheduled in the last timeslot of the day.

$$\sum_{i=1}^{n} z_{ij} = 0, \quad 1 \le j \le 45.$$

2. Penalty for students with three or more consecutive events in a day.

$$\sum_{j=1}^{n} \sum_{l=1}^{m} \sum_{k=a}^{a+2} x_{ijkl} \le 2, \quad 1 \le i \le p;$$

$$a = 1, 2, \ldots, 7, 10, 11, \ldots, 16, \ldots, 37, 38, \ldots, 43.$$

3. Penalty for student with a single event in a day.

$$\sum_{j=1}^{n} \sum_{l=1}^{m} \sum_{k=d}^{d+2} x_{ijkl} > 1, \quad 1 \le i \le p; d = 1, 10, 19, 28, 37.$$

The satisfaction of the following hard constraints is another objective of the UCTP. Their satisfaction is essential for obtaining a meaningful solution, so the violation cannot be accepted at any cost. Any solution satisfying all the hard constraints is considered feasible. The symbol hcv in the following indicates the number of hard constraint violations.

Subject to:

$$\sum_{i=1}^{5} hcv_i$$

1. No student should have more than one event at any timeslot.

$$\sum_{j=1}^{n}\sum_{l=1}^{m} x_{ijkl} \leq 1, \quad 1 \leq i \leq p; \ 1 \leq k \leq 45$$

2. Each event is assigned to a room with enough seats for students and required features.

$$\sum_{i=1}^{45} x_{ijkl} \leq CP(r_l), \quad 1 \leq j \leq n; \ 1 \leq k \leq 45; \ 1 \leq l \leq m; \ \text{and}$$

$$y_{ijk} \leq rf_{jk}, \quad 1 \leq i \leq n; \ 1 \leq j \leq m; \ 1 \leq k \leq q.$$

3. Only one event per room in any timeslot.

$$\sum_{j=1}^{n} x_{ijkl} \leq 1, \quad 1 \leq i \leq p; \ 1 \leq k \leq 45; \ 1 \leq l \leq m.$$

4. Events should only be allocated to timeslots which are predefined as available for them.

$$z_{ij} \leq et_{ij}, \quad \text{where } et_{ij} \in [ET]_{n \times 45}; \ 1 \leq i \leq n; \ 1 \leq j \leq 45.$$

5. Where specified, events should be scheduled to appear in the precise order.

$$\text{if } z_{ik_1} = z_{jk_2} = 1 \wedge ep_{ij} = 1; \text{ then}$$

$$k_1 < k_2, \quad i,j = 1,2,\ldots,n \wedge i \neq j.$$

The objective is to obtain an optimal or near-optimal solution by satisfying all the hard constraints and minimizing soft constraint violations as far as possible. For the sake of simplicity, a direct representation of the solution is taken. A solution involves an integer-valued ordered list of size $|E|$, say $a[i]$ $(1 \leq a[i] \leq 45 \text{ and } 1 \leq i \leq |E|)$. Here, $a[i]$ represents the timeslot for event e_i. A matching algorithm is further used to generate the assignment of rooms. We have a set of events appearing in a timeslot and a list of pre-processed rooms based on their sizes and features. Employing a deterministic network flow algorithm, given by [46], a bipartite matching algorithm is used to get a maximum cardinality match between these two sets. For remaining unplaced events, the bipartite matching algorithm takes them in order and assigns each individual into a suitable room engaged with the least number of events. After applying these procedures, a similar integer-valued ordered list of size $|E|$, say $b[i]$ $(1 \leq b[i] \leq m$ and $1 \leq i \leq |E|)$ is obtained for the event-room assignments. Here, m is the total number of rooms. In the case of a tie, the first room is selected. This results in the complete assignment of all events to appropriate rooms and timeslots.

1.4 ANT COLONY OPTIMIZATION

This section explains the ACO algorithm which is employed to solve the UCTP and establish the proposed ACOWSG algorithm.

ACO is a population-based metaheuristic approach proposed by [37] and is extensively used for solving a number of combinatorial optimization problems. These algorithms have demonstrated their utility in problems that can be reduced to graphs in order to find better paths. The foraging nature of physical ant colonies was the central theme behind the establishment of the ACO algorithm. This algorithm's inspiring source is the pheromone trail leaving and following real ants' behavior, which uses pheromones as an interaction medium. An ant constructs a solution for a given problem in an ACO algorithm, and a population of ants is maintained. During the evolutionary process, useful information is kept as global information (pheromones), which will be updated and used to form individuals' next generation. As a general rule, a generic ACO approach consists of three iterated phases. In the first phase, the formation of a solution by the artificial ants is achieved; in the next step, the pheromone update is performed; and finally, daemon actions like executing an LS are implemented.

The algorithm used in our work is centered on ACO and categorized as an Ant Colony System (ACS), the earliest representative of the ACO class of approaches suggested in [47]. The outline of an ACS for solving the UCTP is presented as Algorithm 1. Here, total m ants are taken, which creates a comprehensive assignment of events to timeslots in each iteration by considering event by event. By selecting the next event from a pre-processed list of events, a single assignment of an event to a timeslot is formed by utilizing an ant and taking a timeslot based on some reasonable probability. Two categories of information used for conducting such procedures are heuristic and stigmergic. The information considering the evaluation of the constraint violations after the assignment is termed as the heuristic information, whereas stigmergic information assesses the effectiveness of making the assignment in the form of a "pheromone" level and is estimated based on the preceding algorithmic moves. To show the stigmergic information, a "pheromone" valued matrix $\omega : E \times T \to R$ with non-negative R values is used. Here T and E denote the sets of timeslots and events, respectively. These pheromone values are initialized to ω_0 before being revised by local and global rules. If an event–timeslot pair is a part of a quality solution, then the pair gets higher chances of repeated selection with a high pheromone value in the future. A matching algorithm is then applied to transform an event–timeslot assignment into a candidate solution at the end of the iterative construction. An LS approach is further used to enhance the quality of the candidate solution (or timetable). The value of the global pheromone is revised using the best quality solution obtained so far during the entire procedure. The pheromone values update process is executed once the candidate solutions are created successfully by all the m ants. The complete procedure of this phase is repeated until the allotted time limit is not exhausted.

The total ordering $\prec$ between the events e and e' is same as defined in [39]. In order to construct an event–timeslot assignment, each ant allocates timeslots to the events sequentially. This is processed as per the order $\prec$. This implies that this process establishes assignments $Z_i : E_i \rightarrow T$ for $i = 0,...,n$, where the set E_i is defined as $E_i := \{e_1, ...,e_i\}$ for the totally ordered events expressed as $e_1 \prec e_2 \prec ... \prec e_n$. Starting with an initial empty assignment $Z_0 = \emptyset$, the assignment Z_i is constructed after the construction of Z_{i-1} as $Z_i = Z_{i-1}$. Here, $t \in T$ and is picked arbitrarily using the subsequent probabilities:

$$p\left(t = t' | Z_{i-1}, \omega\right) = \frac{\omega\left(e_i, t'\right)^{\alpha} \cdot \rho\left(e_i, t'\right)^{\beta} \cdot \psi\left(e_i, t'\right)^{\gamma}}{\sum_{v \in T} \omega\left(e_i, v\right) \cdot \rho\left(e_i, v\right)^{\beta} \cdot \psi\left(e_i, v\right)^{\gamma}} \quad (1.1)$$

Here, the weight of the heuristic information corresponds to hard constraint violations (hcv) and soft constraint violations (scv) are controlled by the parameters β and γ, respectively. Heuristic functions ρ and ψ are given as follows:

$$\rho\left(e_i, t'\right) := \frac{1}{1 + \sum_{e \in Z_{i-1}^{-1}(t')} c\left(e_i, e\right)} \quad (1.2)$$

$$\psi\left(e_i, t'\right) := \frac{1}{1 + L + S + R_{-1} + R_0 + R_1} \quad (1.3)$$

The heuristic function ρ is utilized to assign the higher weight to those timeslots, which produce fewer student clashes. Similarly, the function ψ is used to give more weight to timeslots that produce a lower number of scv. Terms used in the heuristic function ψ can be given as follows:

$$L := \begin{cases} s(e_i) & \text{if the final timeslot of the day is } t', \text{ and} \\ 0 & \text{otherwise} \end{cases} \quad (1.4)$$

S: = # students taking event e_i but on the same day not taking any other events as t' in Z_{i-1},

R_{-1}: = # students taking event e_i and on the same day taking events in the two timeslots earlier to t',

R_0: = # students taking event e_i and on the same day taking events in both, previous and next, timeslots to t',

R_1: = # students taking event e_i and on the same day taking events in the two timeslots later to t'.

Algorithm 1   ACS Alogrithm

$\omega(e, t) \leftarrow \omega_0 \ \forall \ (e, t) \in E \times T$

Input: UCTP instance I;

Output: The best solution s_{best} for I.

1: begin
2:   determine $c(e, e') \ \forall \ (e, e') \in E^2$;   ▷ $c(e, e') :=$ # students attending both e and e'
3:   determine $d(e) \ \forall \ e \in E$;   ▷ $d(e) := |\{e' \in E \setminus \{e\} \mid c(e, e') \neq 0\}|$
4:   determine $f(e) \ \forall \ e \in E$;   ▷ $f(e) :=$ # features requisite by event e
5:   determine $s(e) \ \forall \ e \in E$;   ▷ $s(e) :=$ # students appearing in event e
6:   sort E as per ordered relation $\prec$, providing $e_1 \prec e_2 \prec \ldots \prec e_n$;
7:   $i \leftarrow 0$;   ▷ $i \leftarrow$ existing iteration number
8:   repeat
9:     $i \leftarrow i + 1$;
10:     for $(a \leftarrow 1 \ to \ m)$ do   ▷ construction process of ant a
11:       $Z_0 \leftarrow \phi$;   ▷ $Z_0 \leftarrow$ initial empty assignment
12:       for $(j \leftarrow 1 \ to \ n)$ do   ▷ $j \leftarrow$ event counter
13:         based on the probability distribution p an arbitrary timeslot t is selected for event e_j;
14:         for $\omega(e_j, t)$, local pheromone update is performed;
15:         $Z_j \leftarrow Z_{j-1} \cup (e_j, t)$;
16:       end for
17:       $s \leftarrow$ solution achieved by implementing matching algorithm to Z_n first and then using LS for $h(i)$ steps;
18:       $s_{\text{best}} \leftarrow$ best outcome between s and c_{best}; ▷ $c_{\text{best}} \leftarrow$ current best outcome
19:     end for
20:     update global pheromone using c_{best} for $\omega(e, t) \ \forall \ (e, t) \in E \times T$;
21: until (termination criteria not met);
22: end

An updated rule on the pheromone matrix is executed locally after every construction step. This is performed for the entry corresponds to the current event e_i and the selected timeslot t_{select}. The pheromone decay parameter $\pi \in [0, 1]$ is used to handle the construction process's diversification. Its value is reciprocal to the probability of selecting the same event–timeslot pair in future steps. This implies that the possibility of selecting the same event–timeslot pair will be very high in future steps if the value of the parameter π is minimal.

$$\omega\left(e_i, t_{\text{select}}\right) \leftarrow \left(1-\pi\right) \cdot \omega\left(e_i, t_{\text{select}}\right) + \pi \cdot \omega_0 \tag{1.5}$$

A timetable by means of a candidate solution s is generated by applying a matching algorithm for the appropriate allocation of rooms to event–timeslot pair after the assignment Z_n is completed. The LS approach is then employed for $h(j)$ number of steps over s. The value of $h(j) : j \in N$ depends on the current iteration number j. After each iteration, the pheromone matrix ω is updated globally. First, Z_{best} is treated to be the assignment corresponding to the best solution s_{best} obtained so far. Next, we update ω corresponding to each event–timeslot pair (e,t) as follows:

$$\omega(e,t) \leftarrow \begin{cases} (1-\tau) \cdot \omega(e,t) + \tau \cdot \dfrac{W}{1+q(s_{\text{best}})} & \text{if } Z_{\text{best}}(e) = t, \text{ and} \\ (1-\tau) \cdot \omega(e,t) & \text{otherwise} \end{cases} \tag{1.6}$$

Here, the parameter W controls the pheromone amount laid down by the update rule. The quality of an individual solution s is evaluated by using the function q calculating the fitness function value and defined as follows:

$$q(s) := k \times \text{hcv}(s) + \text{scv}(s)$$

where $hcv(s), \text{scv}(s)$, k are the numbers of hcv and scv on s, and a constant having a large value, generally more than the maximum possible scv, respectively. The function q is used for evaluation of fitness function value and therefore referred to as the fitness function.

An LS is a classical method for finding optimal solutions to many combinatorial optimization problems using two phases. The first phase gives the feasibility, and the second optimizes soft constraints without violating the search space's feasibility. Starting with an empty schedule, the algorithm gradually constructs a timetable by inserting one event into a timetable in the first phase. Usually, the initial timetable is of poor quality, involving many constraint violations. The second phase gradually improves the timetable's quality by altering some of its events to achieve a better timetable. The selection of appropriate neighborhoods constitutes one of the main parts of the LS. These two phases of LS are applied to each individual solution to solve the UCTP. The first phase operates randomly, attempting all probable neighborhood moves for each event from the list of given events associated with hcv while ignoring all scv until the termination condition is met. The pre-specified number of iterations exhausted or an improvement in the solution can be taken as the termination criteria. For simplicity, a portion of the given solution is customized to form a new neighboring solution. In our work, a neighborhood is defined as the union of two smaller neighborhoods, N_1 and N_2, which are defined as follows:

N_1: operator randomly selects a single event and moves this event to a different timeslot, which produces the smallest penalty.

N_2: operator swaps the timeslots of two randomly selected events.

Neighborhood operator N_2 is applied only when N_1 fails. The resulting room allocation disturbance is resolved using the bipartite graph matching algorithm to the affected timeslots after each neighborhood moves with its delta-evaluate measure. A delta-evaluation refers to the computation of the hcv of those events that move within a solution and obtains the fitness function value dispute in the solution between the pre- and post-move of the corresponding event. If there is no new move left in the neighborhood or no hcv for the current event, the first phase continues to the next event. In case there still exists some hcv after exploiting every neighborhood move to all the events, this phase concludes the process by concluding that no potentially viable solution to the problem exists. The second phase starts once we receive a solution from the first phase satisfying all the hard constraints. This phase behaves similar to the first phase by considering soft constraints instead of hard constraints. It tries to reduce the scv by exploiting all the neighborhood moves in order over each event without disturbing the hard constraints. In brief, the first phase provides a feasible solution, and the second phase provides the optimal solution by satisfying as many soft constraints as possible. The LS algorithm's general framework in its construction and improvement phases is depicted in Algorithms 2 and 3.

Algorithm 2   First phase of LS approach

Input: A solution I of UCTP
Output: Either a feasible solution I or no feasible solution
1: begin
2: create a circular randomly-ordered list $(e_1, e_2, \ldots, e_n)$ of events;
3: $i \leftarrow 0$;   ▷ $i \leftarrow$ event counter
4: select event e_i after $i \leftarrow i + 1$;   ▷ pointer moves to next event
5: if (every neighborhood moves exploited over every events) then
6:     if ($\exists$ any hcv in I) then
7:         STOP LS with no feasible solution;
8:     else
9:         END the first phase with a feasible solution I as output;
10:    end if
11: end if
12: if ((feasible e_i) $\bigvee$ (no untried move left for e_i)) then
13:     goto 4;
14: end if
15: $Examine(e_i, I)$;   ▷ apply all neighborhood moves and return the solution I
16: if (# hcv reduced in I) then
17:     consider the move;
18:     goto 3;
19: else
20:     goto 12;
21: end if
22: end

Algorithm 3   Second phase of LS approach

Input: A solution I without hcv from Algorithm 2
Output: A possibly improved solution I
1: begin
2: use circular randomly-ordered list of n events generated in Algorithm 2;
3: $i \leftarrow 0$;
4: select event e_i after $i \leftarrow i + 1$;
5: if (every neighborhood moves exploited over every events) then
6:     STOP LS with a best solution I;
7: end if
8: if ((e_i NOT associated in any scv) $\bigvee$ (no untried move left for e_i)) then
9:     goto 4;
10: end if
11: $Examine(e_i, I)$;   ▷ apply all neighborhood moves and return the solution I
12: if (# scv reduced without violating the feasibility of I) then
13:     apply the move;
14:     goto 3;
15: else
16:     goto 8;
17: end if
18: end

Procedure $Examine(e_i, I)$

Solution I with event e_i are arguments. Returns I after applying neighborhood moves

Input: Set T of 45 timeslots, and set R of of m rooms

1: begin
2:   apply N_1 to solution I;
3:   if (N_1 effective) then
4:     produce solution I;
5:   else
6:     employ N_2 to I and produce solution I;
7:   end if
8:   for ($k \leftarrow 1$ to 45) do
9:     if (timeslot t_k is effected by either of N_1 and N_2) then
10:       use the matching algorithm for events held in t_k to allocate rooms ;
11:     end if
12:   end for
13:   delta-evaluate the result of the move;
14:   return I;
15: end

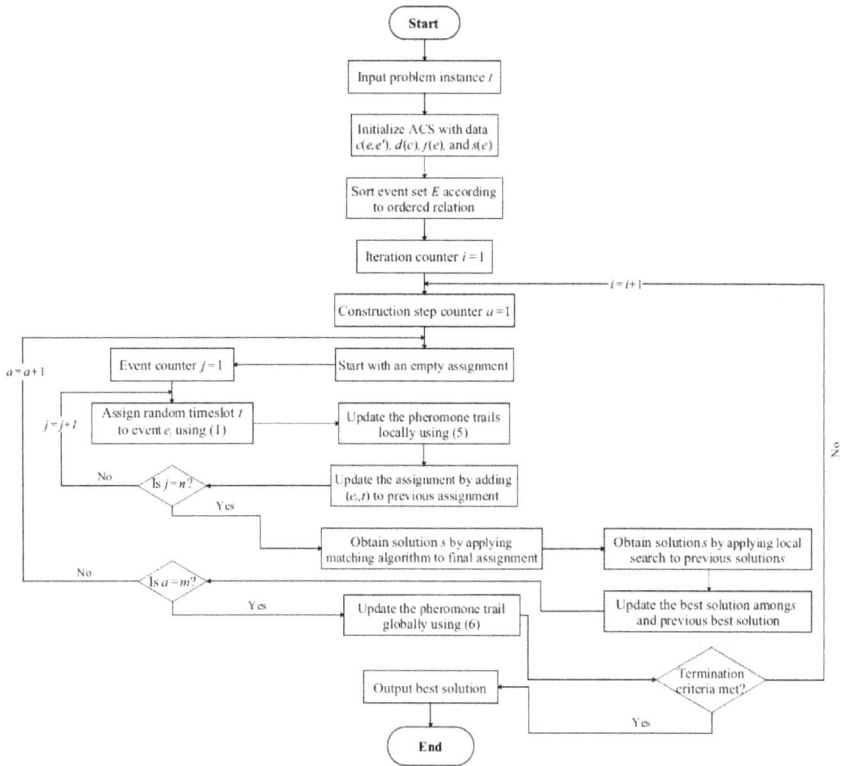

Figure 1.1 Flowchart of ACS algorithm.

The complete working procedure of ACS algorithm is illustrated by the following flowchart given in Figure 1.1.

1.5 THE PROPOSED ALGORITHM

The developed ACOWSG algorithm by hybridizing an ACO approach with student groupings for the solution of the UCTP is described in the present section. Reference [44] defines the definition of grouping problem as the one where the task is to segregate an object set X into a collection of mutually disjoint subsets $x_i \in X$ such that $x_i \cap x_j = \phi$ for $i \neq j$ and $\cup x_i = X$, and according to a set of problem-specific constraints defining valid and legal groupings. Based on the notion of grouping, [44] proposed the grouping GA to solve the clustering problems. This grouping approach is not a ready-for-use method and must be tailored distinctly according to the problem. In our work, we have exploited this approach to construct groups of students. Then the events attended by these groups of students are assigned to suitable rooms and timeslots by the ACO algorithm. This approach leads to the innovative and novel idea of using a grouping approach.

In our implementation, a set of students S is considered as the object set and segregated into k mutually disjoint groups $G_i (i = 1,2,...,k)$ in such a manner that $G_i \cap G_j = \phi$ for $i \neq j$ and $\cup G_i = S$. For all the events in the given order, mutually disjoint groups are formed from S. Commencing from e_1, a group is created of those students opting e_1. Any student is excluded from other selections once included in a group. This procedure moves to the next event and continues until all the students are not selected for any group. Presume that $G_1, G_2,...,G_k, k \in \mathbb{I}$ groups are formed and that an event group $E(G_k)$ is created for each group. Here $E(G_k) = \cup_j \{E(s_j) : s_j \in G_k\}$ is the union of all students' event of G_k. Furthermore, the cardinality of G_k is denoted as $|E(G_k)|$.

Next, $|E(G_k)| \forall k$ corresponding to G_k is reduced within a specific bound once the formation of student groups and corresponding event groups are completed. Initially, this upper bound is fixed to 45, that is, the number of timeslots. The restrictions employed by the second and third soft constraints reduced this bound to 30, which can take the maximum number of events per student as the least upper bound. To decrease the group cardinality $|E(G_k)| \forall k$, each G_k is further partitioned into distinct subgroups using the aforementioned grouping process to bring down the group cardinality. The procedure is repeated until $|E(G_k)| \forall k$ is not within the selected limit. Room features and event features are next modified to comply with the second hard constraint. The modified UCTP instances are now solved using the already-discussed ACS algorithm, given as Algorithm 1, by taking each student group $G_i (i = 1,2,...,k)$ as a student. Clearly, $k \leq p$. The initial upper limit for each group's cardinality is taken as 30 and further reduced successively by 1 until the optimal solution is not achieved, or the upper limit becomes equivalent to the maximum number of events per student. As we know that UCTP is a multidimensional assignment problem assigning events into timeslots and rooms in such a manner that there should not be any hcv, and the penalty value of scv should be minimum. Hence, events taken by student groups, instead of students, are assigned to timeslots and suitable rooms.

Algorithm 4   Student grouping algorithm

Input: Student set S; event set E; event set $E(s_i)$ taken by student s_i
Output: SG: groups of students with corresponding group of events within required
   cardinality
1: **begin**
2:   $M \leftarrow \max\limits_{1 \leq i \leq p} \{|E(s_i)|\}$;
3:   MPC $\leftarrow$ (maximum possible cardinality of group | $M \leq$ MPC $\leq |T|$); $\triangleright T$ is set of
   timeslots
4:   $S' \leftarrow S$;
5:   $i \leftarrow 1$;
6:   $k \leftarrow 1$;
7:   SG $\leftarrow \phi$; $\triangleright$ SG$\leftarrow$ groups of students taking group of students, say G_k, as elements
8:   **while** $(S' \neq \phi)$ **do**
9:     $G_k \leftarrow \phi$; $\triangleright$ $G_k \leftarrow$ set of kth group consisting of students as elements
10:     $E(G_k) \leftarrow \phi$; $\triangleright$ $E(G_k) \leftarrow$ set consisting of events taken by the kth group
11:     **for** $(j \leftarrow 1 \ to \ p)$ **do**
12:       **if** $((e_i \in E(s_j)) \bigwedge (s_j \in S'))$ **then**
13:         $G_k \leftarrow G_k \bigcup \{s_j\}$;
14:         $E(G_k) \leftarrow E(G_k) \bigcup E(s_j)$;
15:         $S' \leftarrow S' \setminus \{s_j\}$;
16:       **end if**
17:     **end for**
18:     **if** $(G_k \neq \phi)$ **then**
19:       $SizeCheck(e_i, G_k, E(G_k), \text{MPC})$; $\triangleright$ returns the group of students 'Group'
   with required cardinality
20:       SG $\leftarrow$ SG $\bigcup Group$;
21:       $k \leftarrow k + 1$;
22:     **end if**
23:     $i \leftarrow i + 1$;
24:   **end while**
25: **end**

Procedure $SizeCheck(e_i, G_k, E(G_k), \text{MPC})$

$e_i, G_k, E(G_k)$ and MPC are taken as arguments; the procedure returns the group of students 'Group'
within the required cardinality
1: **begin**
2:   $Group \leftarrow \phi$; $\triangleright$ $Group \leftarrow$ a set consisting of students subgroup of G_k within required cardinality
3:   **if** $(|E(G_k)| \leq \text{MPC})$ **then**
4:     $Group \leftarrow \{G_k\}$;
5:     **return** $Group$;
6:   **else**
7:     **repeat**
8:       $GroupSizeReduced(e_i, G_k, E(G_k))$; $\triangleright$ returns a subgroup A_k^c of G_k with reduced
   cardinality
9:       $E(A_k^c) \leftarrow \bigcup_y \{E(s_y)| \ s_y \in A_k^c\}$;
10:       **if** $(|E(A_k^c)| \leq \text{MPC})$ **then**
11:         $Group \leftarrow Group \bigcup \{A_k^c\}$;
12:       **end if**
13:     **until** $(|E(A_k^c)| > \text{MPC} \ \ \text{for any} \ A_k^c \subseteq G_k \ | \ \bigcup_c (A_k^c) = G_k)$;
14:     **return** $Group$;
15:   **end if**
16: **end**

Procedure $GroupSizeReduced(e_i, G_k, E(G_k))$

e_i, G_K and $E(G_k)$ are taken as arguments; the procedure returns several partitioned subgroups A_k^c of G_k

```
 1: begin
 2:   c ← 1;
 3:   for (j ← i + 1 to n) do
 4:     if ((e_j ∉ ∩_y{E(s_y)|s_y ∈ G_k}) ⋀(G_k ≠ φ)) then
 5:       A_k^c ← φ;
 6:       E(A_k^c) ← φ;
 7:       for ((l ← 1 to p) ⋀(s_l ∈ G_k) ⋀(e_j ∈ E(s_l))) do
 8:         A_k^c ← A_k^c ⋃{s_l};
 9:         E(A_k^c) ← E(A_k^c) ⋃ E(s_l);
10:         G_k ← G_k ∖ {s_l};
11:       end for
12:       if (A_k^c ≠ φ) then
13:         return A_k^c;
14:         c ← c + 1;
15:       end if
16:       i ← i + 1;
17:     end if
18:   end for
19: end
```

After generating disjoint groups of students, room features and event features are next updated to fulfill the constraints' room-capacity and room-feature requirements. In the beginning, there are q features for each of the m rooms, assuming that multiple rooms have identical seating capacities. Let there be only x distinct capacities among these rooms. An additional x feature will be then added to each room with the existing features. Similar-capacity rooms will get a similar set of new features. The following procedure returns the modified room features $f[i][q+x]$ for room r_i.

Procedure Modified room-features

Input: Room set R of m rooms along with their capacities and features
Output: Modified room-features

```
 1: begin
 2:   for (i ← 1 to m) do
 3:     cp[i] ← capacity of r_i;
 4:   end for
 5:   distinct room capacities are sorted in their increasing order;
 6:   c[1] ← smallest room capacity;
 7:   c[x] ← largest room capacity;                          ▷ x distinct room capacities are assumed
 8:   f[i][q] ← existing q features of r_i;
 9:   for (i ← 1 to m) do                                    ▷ for m rooms
10:     take existing room-features f[i][q] of r_i;
11:     for (j ← 1 to x) do                                  ▷ for distinct x room capacities
12:       if (cp[i] ≤ c[j]) then
13:         f[i][q + j] ← 1;
14:       else
15:         f[i][q + j] ← 0;
16:       end if
17:     end for
18:   end for
19: end
```

Similarly, the event features are modified for the new problem instance. In this case, the total number of students enrolled in every event is an essential requirement. The subsequent procedure will give us the modified event features $g[i][q+x]$ for event e_i.

Procedure Modified event-features

Input: Room set R of m rooms along with their capacities; event-features for all n events; students enrolled in each event

Output: Modified event-features

1: begin
2: $c[0] \leftarrow 0$;
3: distinct room capacities are sorted in their increasing order;
4: $c[1] \leftarrow$ smallest room capacity;
5: $c[x] \leftarrow$ largest room capacity;
6: $e[i] \leftarrow$ total students enrolled in e_i;
7: $g[i][q] \leftarrow$ existing q features of e_i;
8: **for** $(i \leftarrow 1 \ to \ n)$ **do**   ▷ for n events
9:    take existing event-features $g[i][q]$ of e_i;
10:    **for** $(j \leftarrow 1 \ to \ x)$ **do**
11:      **if** $(c[j-1] < e[i] \leq c[j])$ **then**
12:        $g[i][q+j] \leftarrow 1$;
13:      **else**
14:        $g[i][q+j] \leftarrow 0$;
15:      **end if**
16:    **end for**
17: **end for**
18: **end**

Algorithm 5 describes the proposed ACOWSG algorithm. The algorithm's termination criteria are the completion of the computation time or completion of the number of iterations or the attainment of the optimal solution.

Algorithm 5 Proposed ACOWSG Algorithm

Input: UCTP instance I;

Output: Either a solution s_{best} or no solution.

1: begin
2: enforce 'Student grouping algorithm' with MPC as an upper bound for group's cardinality;   ▷ given in Algorithm 4.
3: modified UCTP instance after employing event-features and room-features procedures;
4: apply ACS algorithm   ▷ given in Algorithm 1
5: **if** (termination criteria met) **then**
6:    STOP the algorithm;
7: **else**
8:    MPC $\leftarrow$ MPC $- 1$;   ▷ the upper bound of group cardinality is reduced by one
9:    goto 2;
10: **end if**
11: **end**

1.6 EXPERIMENTAL RESULTS

We performed the experiments on a 3.10 GHz PC with 2 GB of random-access memory. All the algorithms were coded in GNU C++ under version 4.5.2. The proposed ACOWSG algorithm's performance is investigated by comparing its results with those obtained from other existing algorithms in terms of the fitness function value. Two different sets of problem instances are considered for this purpose. The first benchmark of problem instances is considered from [38], whereas the second set is taken from the second track of ITC2007. During the experiment, the group cardinality is set moderately higher than the maximum number of events per student in both sets of problem instances. Although many real-world objects and constraints seem to be absent from these considered problem instances, they permit us to compare the proposed approach with those existing algorithms. These different sets of problem instances are dealt with separately in further subsections.

1.6.1 Experiments on Socha's benchmark dataset

The detailed description of the first set of 11 benchmark problem instances is displayed in Table 1.1. These problem instances are classified into small (s), medium (m), and large (l) classes. In the table, n,m,q, and p indicate the total number of events, rooms, room features, and students, respectively. Also, $S/E, E/S, F/R$, and F/E stand for students per event, events per student, features per room, and features per event, respectively.

A comprehensive experiment was performed over several parameter combinations to identify the optimal set of parameters for the ACOWSG algorithm. The outcomes are consistent with those suggested by [39] and illustrated in Table 1.2.

Table 1.1 Description of [38] Problem Instances

Instance	n	m	q	p	Max SE	Max ES	Avg. FR	Avg. FE
s01	100	5	5	80	15	15	2.8	1.88
s02	100	5	5	80	13	17	3.0	2.02
s03	100	5	5	80	20	13	3.0	2.21
s04	100	5	5	80	12	12	4.4	2.92
s05	100	5	5	80	17	19	3.8	2.80
m01	400	10	5	200	11	20	2.9	2.355
m02	400	10	5	200	11	20	3.0	2.33
m03	400	10	5	200	12	20	3.2	2.525
m04	400	10	5	200	11	20	3.1	2.493
m05	400	10	5	200	20	20	3.2	2.535
ℓ	400	10	10	400	30	20	4.8	4.37

Table 1.2 Selected Parameters for ACOWSG Algorithm

Parameter	Small	Medium	Large
m	15	15	10
ω_0	0.5	10	10
τ	0.1	0.1	0.1
α	1	1	1
β	3	3	3
γ	2	2	2
π	0.1	0.1	0.1
$h(j)$	$\begin{cases} 5000 & \text{for } j=1 \\ 2000 & \text{for } j \geq 2 \end{cases}$	$\begin{cases} 50000 & \text{for } j \leq 10 \\ 10000 & \text{for } j \geq 11 \end{cases}$	$\begin{cases} 150000 & \text{for } j \leq 20 \\ 100000 & \text{for } j \geq 21 \end{cases}$
W	10^5	10^{10}	10^{10}

Table 1.3 f_{min} by ACO and ACOWSG for [38] Instances

Approach	s01	s02	s03	s04	s05	m01	m02	m03	m04	m05	—
ACO	3	9	6	1	0	250	244	334	225	236	∞
Time	2.173	2.107	2.287	1.741	2.107	111.38	118.49	119.70	115.06	117.01	—
ACOWSG	0	0	0	0	0	209	202	276	200	211	1021
Time	0.464	0.392	0.260	0.472	0.424	119.65	114.41	112.72	107.71	112.53	599.01

The proposed ACOWSG algorithm is first compared with the ACO algorithm. For this objective, all of these 11 problem instances are run for 50 independent trials, and the least value of fitness function (f_{min}) among them is taken as the optimum solution. The run time for each small, medium, and large problem instances are restricted to 2, 120, and 600 seconds, respectively. Similarly, 200, 100000, and 100000 are the maximum numbers of iterations in the local search for these three classes. Either consummation of the computation time or completion of the number of iterations, or attainment of the optimal solution is used as the algorithm's termination criteria. Table 1.3 gives a synopsis of the comparison of ACO and ACOWSG algorithms in terms of f_{min}. From this table, we can clearly observe that the proposed algorithm performs better than the ACO algorithm. Notably, the ACO is incapable of producing a feasible solution for the large-sized instance, whereas the proposed algorithm can produce the same within the time limit.

This comparison result is also demonstrated by plotting f_{min} versus time graphs for the same set of problem instances in Figures 1.2–1.4. The first part of Figure 1.4 shows the total number of hcv, that is, distance to feasibility, for the best solution obtained by ACO and ACOWSG. The second part shows the scv obtained by ACOWSG.

Next, we compare the performance of the proposed algorithm with that of existing state-of-the-art approaches. The computation time limit is expanded to 900 and 9,000 seconds, respectively, for medium and large instances, a frequently used time limit available in the literature. The number of independent trials for large and medium instances is also fixed to 30 and 60, respectively. The outcome for small-sized UCTP instances, displayed in Table 1.3, are already optimal and considered the same for this purpose. In Table 1.4, the performance of the ACOWSG algorithm is displayed in terms of the best solution (f_{min}), the worst solution (f_{max}), the average (f_{avg}), the standard deviation (s), and the median (f_{med}) among all the obtained outcomes. The table also demonstrates the duration (time) in seconds to get f_{min}. These results are next presented as graphs for all the small, medium, and large problem instances in Figure 1.5. Here x and y axes denote time (in seconds) and f_{max}, respectively.

We compare the performance of the proposed ACOWSG algorithm with 18 existing algorithms in the literature. All these considered algorithms are described in Table 1.5. Here, we want to emphasize that the compared algorithms' prerequisites beneath which their outcomes were stated to have been frequently utilized to compare the proposed algorithms' performance. However, the methodology is not entirely unbiased as the conditions may differ from other approaches, yet the reported results can give us a

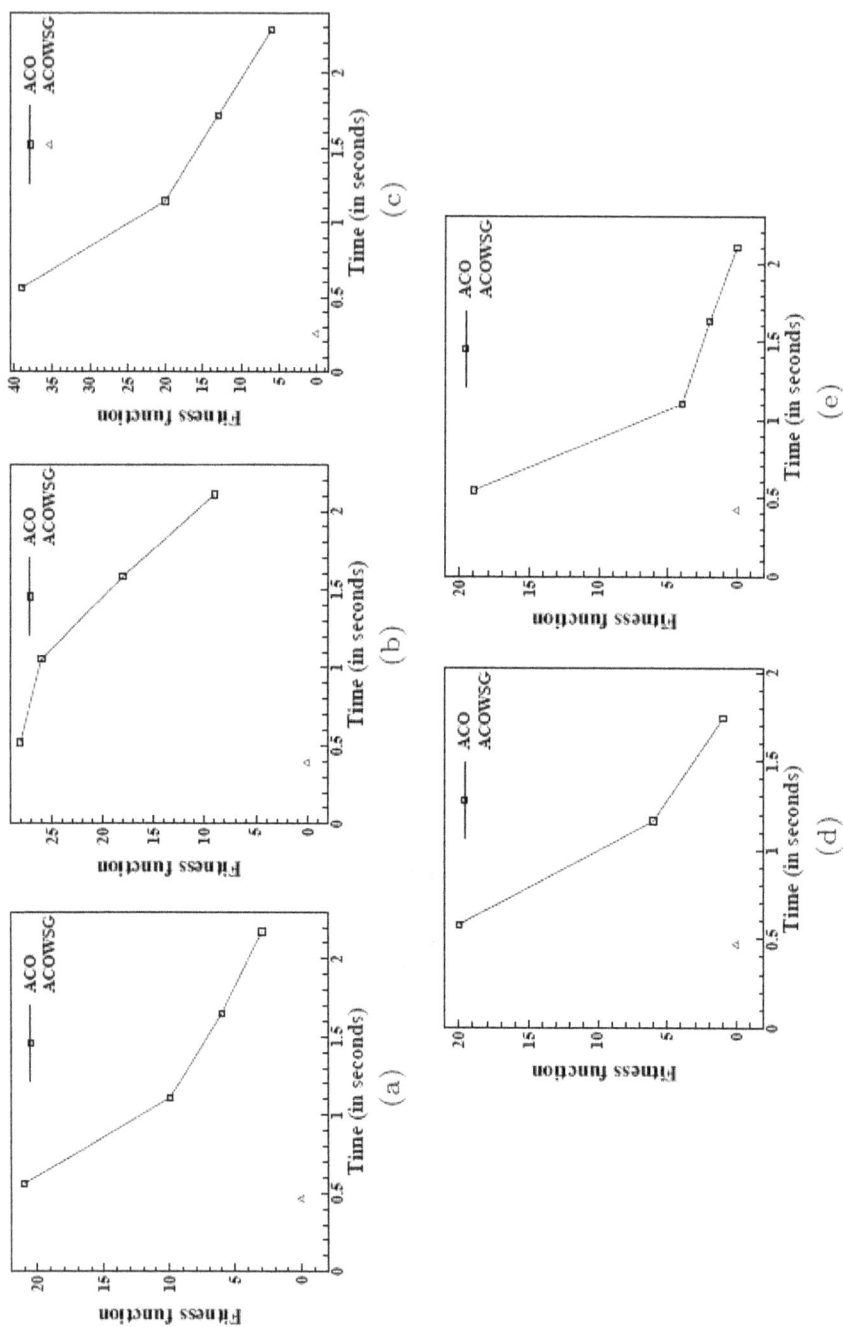

Figure 1.2 f_{min} versus time for (a) s01, (b) s02, (c) s03, (d) s04, and (e) s05 UCTP instances by ACO and ACOWSG.

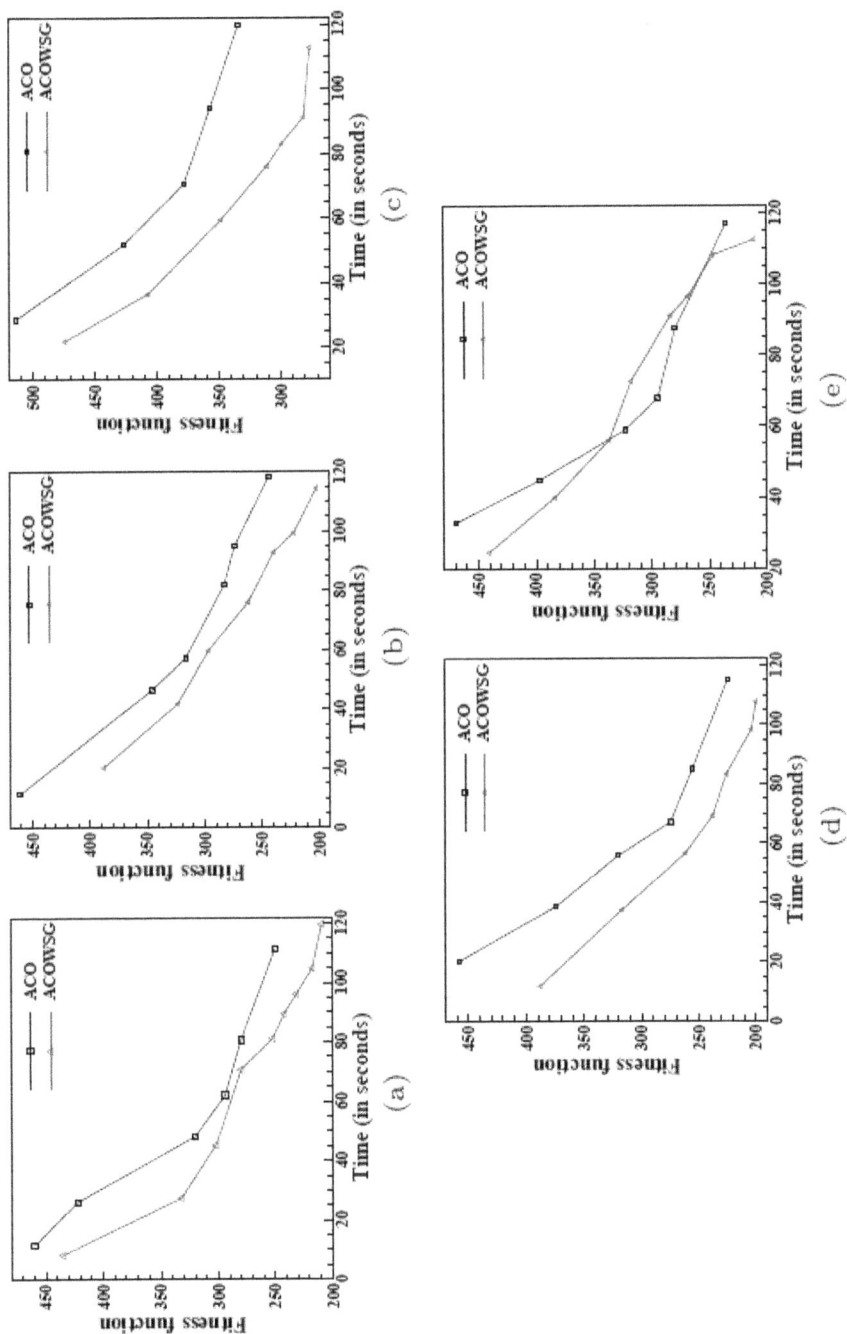

Figure 1.3 f_{min} versus time for (a) m01, (b) m02, (c) m03, (d) m04, and (e) m05 UCTP instances by ACO and ACOWSG.

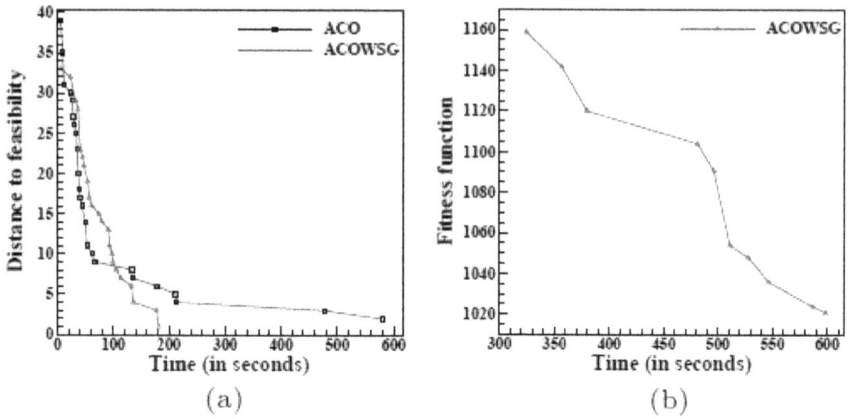

Figure 1.4 (a) Distance to feasibility versus time and (b) f_{min} versus time by ACO and ACOWSG for large problem instance.

Table 1.4 Output of [38] Problem Instances

Instance	f_{min}	f_{max}	f_{avg}	σ	f_{med}	Time (in seconds)
s01	0	0	0	0	0	0.464
s02	0	0	0	0	0	0.392
s03	0	0	0	0	0	0.260
s04	0	0	0	0	0	0.472
s05	0	0	0	0	0	0.424
m01	109	152	127.5	13.35	124.5	876.20
m02	111	153	131.4	14.16	130	888.30
m03	129	205	159.67	21.67	155.5	846.40
m04	104	136	116.1	10.31	113.5	780.68
m05	110	162	132.23	16.90	130	822.58
ℓ	603	796	669.60	63.84	644	8,687.54

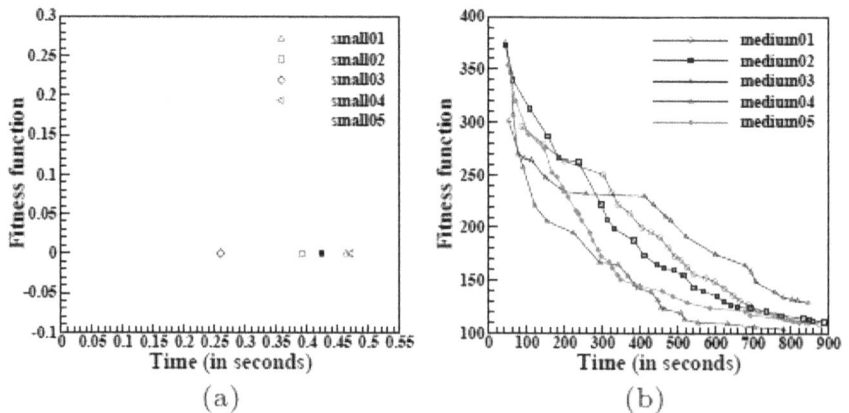

Figure 1.5 f_{min} versus time for (a) small, (b) medium, and (c) large problem instances by ACOWSG.

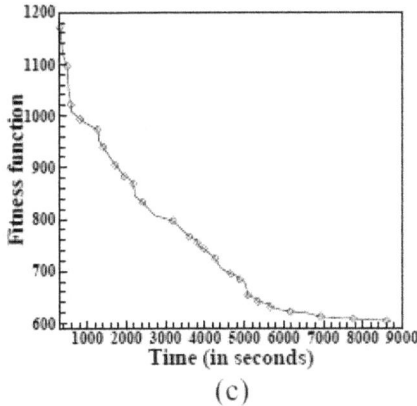

Time (in seconds)

(c)

Figure 1.5 (Continued)

Table 1.5 Keys of the Algorithms Used for Comparison

No.	Key	Used methodology	Reference
1	ACOWSG	Ant colony optimization with student grouping	Proposed
2	A1	GA with a repair function and LS	[10]
3	A2	Composite neighborhood structure with randomized iterative improvement algorithm	[48]
4	A3	TS with graph-based hyper-heuristic	[49]
5	A4	Evolutionary algorithm using a light mutation operator followed by a randomized iterative improvement algorithm	[50]
6	A5	Modified great deluge algorithm by using a nonlinear decay of water level	[27]
7	A6	MAX-MIN ant system using a separate LS routine	[38]
8	A7	Heuristic approach utilizing fuzzy multiple heuristic	[51]
9	A8	GA with an LS method	[39]
10	A9	ACO algorithm	[39]
11	A10	Electromagnetic-like mechanism and the great deluge algorithm	[52]
12	A11	An average late acceptance randomized descent algorithm	[53]
13	A12	Evolutionary non-linear great deluge which is extension of a non-linear great deluge algorithm	[54]
14	A13	Harmony search algorithm with multi-pitch adjusting rate	[35]
15	A14	Hybridization of SA with ACO	[41]
16	A15	Hybridization of TS with ACS	[41]
17	A16	A new hybrid algorithm combining GA with LS and using events based on groupings of students	[45]
18	A17	A non-dominated sorting GA (NSGA-II) hybridized with two LS technique and a TS heuristic	[31]
19	A18	A hybrid algorithm based on the improved parallel GA and LS	[55]

Table 1.6 First Comparison Results

Instance	ACOWSG	A1	A2	A3	A4	A5	A6	A7	A8	A9	A10	A11	A12	A13	A14	A15	A16	A17	A18
s01	0	2	0	6	0	3	1	10	0	0	0	0	0	0	0	0	0	0	0
s02	0	4	0	7	0	4	3	9	3	0	0	0	1	0	0	0	0	0	0
s03	0	2	0	3	0	6	1	7	0	0	0	0	0	0	0	0	0	0	0
s04	0	0	0	3	0	6	1	17	0	0	0	0	0	0	0	0	0	0	0
s05	0	4	0	4	0	0	0	7	0	0	0	0	0	0	0	0	0	0	0
m01	109	254	242	372	221	140	195	243	280	225	175	143	126	124	117	150	106	127	84
m02	111	258	161	419	147	130	184	325	188	237	197	130	123	117	121	179	107	122	99
m03	129	251	265	359	246	189	248	249	249	292	216	183	185	190	158	183	132	172	142
m04	104	321	181	348	165	112	164.5	285	247	216	149	133	116	132	124	140	72	110	84
m05	110	276	151	171	130	141	219.5	132	232	243	190	169	129	73	134	152	107	160	112
ℓ	603	1026	100% Inf.	1068	529	876	851.5	1138	100% Inf.	100% Inf.	912	825	821	424	645	750	505	904	516

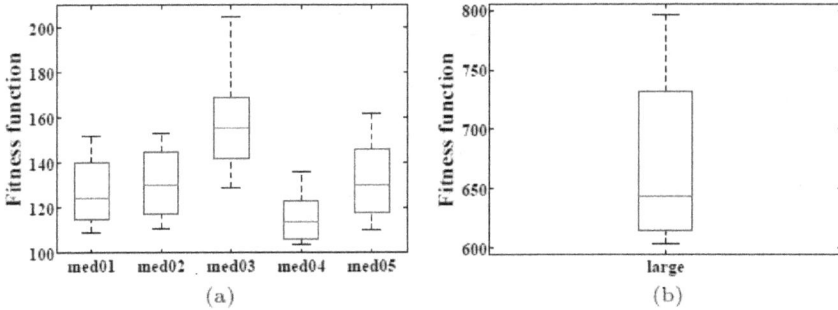

Figure 1.6 Boxplots of results obtained for (a) medium and (b) large UCTP instances.

rough evaluation of the proposed algorithm's quality. Results displayed in Table 1.6 show that the proposed algorithm produces the best outcomes for all the small and medium UCTP instances. Here, x% Inf. means the percentage of runs that failed to find a feasible solution. Boxplots compiling the obtained results from all the independent trials of each medium and large problem instance are given in Figure 1.6. Here the box shows the range of the data between 25% and 75% quantile, whereas the used bar represents the median.

1.6.2 Experiments on ITC2007 benchmark dataset

In the second experiment, the ACOWSG algorithm is applied to solve the UCTP defined in the second track of ITC2007. The problem comprises all the five hard and three soft constraints described in Section 1.3. All the 24 benchmark problem instances of the second track of ITC2007[3] are used to examine the algorithm. The characteristics of the considered problem instances are given in Table 1.7. In this table, the used symbols have their usual meaning.

Thirty independent trials are performed for each problem instance by taking the parameter settings used for the large problem instance in the previous subsection. The outcomes of the experiments for all the 24 problem instances are illustrated in Table 1.8 in terms of f_{min}, f_{max}, f_{avg}, σ, f_{med}, and time (in seconds). The table demonstrates that the ACOWSG algorithm can provide a feasible solution for all the problem instances.

Next, the proposed algorithm is compared with several state-of-the-art algorithms using the same dataset and known and available to the authors. This offers a total of 13 algorithms for comparison, whose descriptions are

---

[3] www.cs.qub.ac.uk/itc2007//Login/SecretPage.php

Table 1.7 Description of Problem Instances Proposed in ITC2007

Instance	n	m	q	p	Max SE	Max ES	Avg FR	Avg FE
2007 – 1	400	10	10	500	33	25	3	1
2007 – 2	400	10	10	500	32	24	4	2
2007 – 3	200	20	10	1000	98	15	3	2
2007 – 4	200	20	10	1000	82	15	3	2
2007 – 5	400	20	20	300	19	23	2	1
2007 – 6	400	20	20	300	20	24	3	2
2007 – 7	200	20	20	500	43	15	5	3
2007 – 8	200	20	20	500	39	15	4	3
2007 – 9	400	10	20	500	34	24	3	1
2007 – 10	400	10	20	500	32	23	3	2
2007 – 11	200	10	10	1000	88	15	3	1
2007 – 12	200	10	10	1000	81	15	4	23
2007 – 13	400	20	10	300	20	24	2	1
2007 – 14	400	20	10	300	20	24	3	1
2007 – 15	200	10	20	500	41	15	2	3
2007 – 16	200	10	20	500	40	15	5	3
2007 – 17	100	10	10	500	195	23	4	2
2007 – 18	200	10	10	500	65	23	4	2
2007 – 19	300	10	10	1000	55	14	3	1
2007 – 20	400	10	10	1000	40	15	3	1
2007 – 21	500	20	20	300	16	23	3	1
2007 – 22	600	20	20	500	22	25	3	2
2007 – 23	400	20	30	1000	69	24	5	3
2007 – 24	400	20	30	1000	41	15	5	3

Table 1.8 Output of Problem Instances of ITC2007

Instance	f_{min}	f_{max}	f_{avg}	σ	f_{med}	Time (in seconds)
2007 – 1	10	211	74.3	87.56	20.5	404.39
2007 – 2	6	354	65.5	108.06	12	385.43
2007 – 3	145	371	195.9	66.10	184	311.1
2007 – 4	258	345	301.4	31.81	294.5	416.91
2007 – 5	0	3	1.4	1.506	1	322.65
2007 – 6	0	0	0	0	0	401.72
2007 – 7	4	7	5.9	1.100	6	360.38
2007 – 8	0	0	0	0	0	12.68
2007 – 9	1	723	181.5	293.80	8.5	294.36
2007 – 10	2	509	102.5	162.86	2	346.14
2007 – 11	15	213	149.7	73.27	175.5	419.42
2007 – 12	0	256	35.2	79.03	0	349.58
2007 – 13	0	156	64.3	69.49	35	310.16
2007 – 14	0	0	0	0	0	178.82
2007 – 15	0	0	0	0	0	132.47
2007 – 16	1	13	6.3	5.100	4	388.68
2007 – 17	0	9	2.8	3.293	1.5	284.67
2007 – 18	0	0	0	0	0	67.59

Table 1.8 (Continued) Output of Problem Instances of ITC2007

Instance	f_{min}	f_{max}	f_{avg}	σ	f_{med}	Time (in seconds)
2007 – 19	710	1448	1078.6	282.08	1097	398.72
2007 – 20	379	449	423.7	20.58	423	406.9
2007 – 21	0	9	2.1	2.923	1	300.2
2007 – 22	7	41	17.9	11.90	16	405.38
2007 – 23	883	1294	1084	156.41	1121	418.73
2007 – 24	3	25	12.2	7.871	10	385.26

Table 1.9 Keys of the Algorithms Used for Comparison

No.	Key	Algorithm	Reference
1	ACOWSG	Ant colony optimization with student grouping	Proposed
2	B1	Hybrid GA and TS approach	[24]
3	B2	Three stage metaheuristic-based algorithm	[56]
4	B3	SA-coloring techniques	[14]
5	B4	Deterministic integer programming based heuristic algorithm	[17]
6	B5	Combination of a general-purpose constraint satisfaction solver, TS, and ILS techniques	[13]
7	B6	Mixed metaheuristic approach including TS and SA used in conjunction with various neighborhood operators	[57]
8	B7	Heuristic approach based on stochastic LS and consisting of several modules	[58]
9	B8	ACO algorithm used in conjunction with a local improvement search routine	[34]
10	B9	LS-based algorithm using routines taken from the constraint solver library	[59]
11	B10	Metaheuristic approach based on SA	[60]
12	B11	Approach combining a generic modeling approach with an adaptive search methodology incorporating learning mechanisms and effective move operators	[61]
13	B12	A hybrid algorithm based on the improved parallel GA and LS	[55]
14	B13	Iterated variable neighborhood descent algorithm	[62]

given in Table 1.9. The proposed algorithm gives an optimal outcome with no *hcv* and *scv* for ten of these UCTP instances. The algorithm also provides improved or equal quality results for 13 of these instances.

In Table 1.10, the term "–" signifies an untried instance in the experiment. Results obtained for all the 24 problem instances from all the 30 independent trials are also compiled by boxplots in Figure 1.7. Here, outliers are indicated as a plus.

Table 1.10 Second Comparison Results

Instance	ACOWSG	B1	B2	B3	B4	B5	B6	B7	B8	B9	B10	B11	B12	B13
2007 – 1	10	501	1166	15	1636	61	571	1482	15	1861	59	650	409	677
2007 – 2	6	342	1665	9	1634	547	993	1635	0	100% Inf.	0	470	381	450
2007 – 3	145	3770	251	174	355	382	164	288	391	272	148	290	195	288
2007 – 4	258	234	424	249	644	529	310	385	239	425	25	600	211	570
2007 – 5	0	0	47	0	525	5	5	229	34	8	0	35	0	30
2007 – 6	0	0	412	0	640	0	0	851	87	28	0	20	0	10
2007 – 7	4	0	6	1	0	0	6	10	0	13	0	30	0	15
2007 – 8	0	0	85	0	241	0	0	0	4	6	0	0	0	0
2007 – 9	1	989	1819	29	1889	0	1560	1947	0	100% Inf.	0	630	0	620
2007 – 10	2	499	2091	2	1677	0	1650	1741	0	100% Inf.	3	2349	476	1764
2007 – 11	15	246	288	178	615	548	178	240	547	263	142	350	135	250
2007 – 12	0	172	474	14	528	869	146	475	32	804	267	480	153	450
2007 – 13	0	0	298	0	485	0	0	675	166	285	1	46	0	30
2007 – 14	0	0	127	0	739	0	1	864	0	110	0	80	0	68
2007 – 15	0	0	108	0	330	379	0	0	0	5	0	0	0	30
2007 – 16	1	0	138	1	260	191	2	1	41	132	0	0	0	20
2007 – 17	0	0	0	–	35	1	0	5	68	72	0	0	0	0
2007 – 18	0	0	25	–	503	0	0	3	26	70	0	20	0	10
2007 – 19	710	84	2146	–	963	100% Inf.	1824	1868	22	100% Inf.	0	360	75	299
2007 – 20	379	297	625	–	1229	1215	445	596	100% Inf.	878	543	150	295	150
2007 – 21	0	0	308	–	670	0	0	602	33	40	5	150	0	0
2007 – 22	7	1142	100% Inf.	–	1956	0	29	1364	0	889	5	33	533	15
2007 – 23	883	963	3101	–	2368	430	238	688	100% Inf.	436	1292	1007	856	892
2007 – 24	3	274	841	–	945	720	21	822	30	372	0	0	266	0

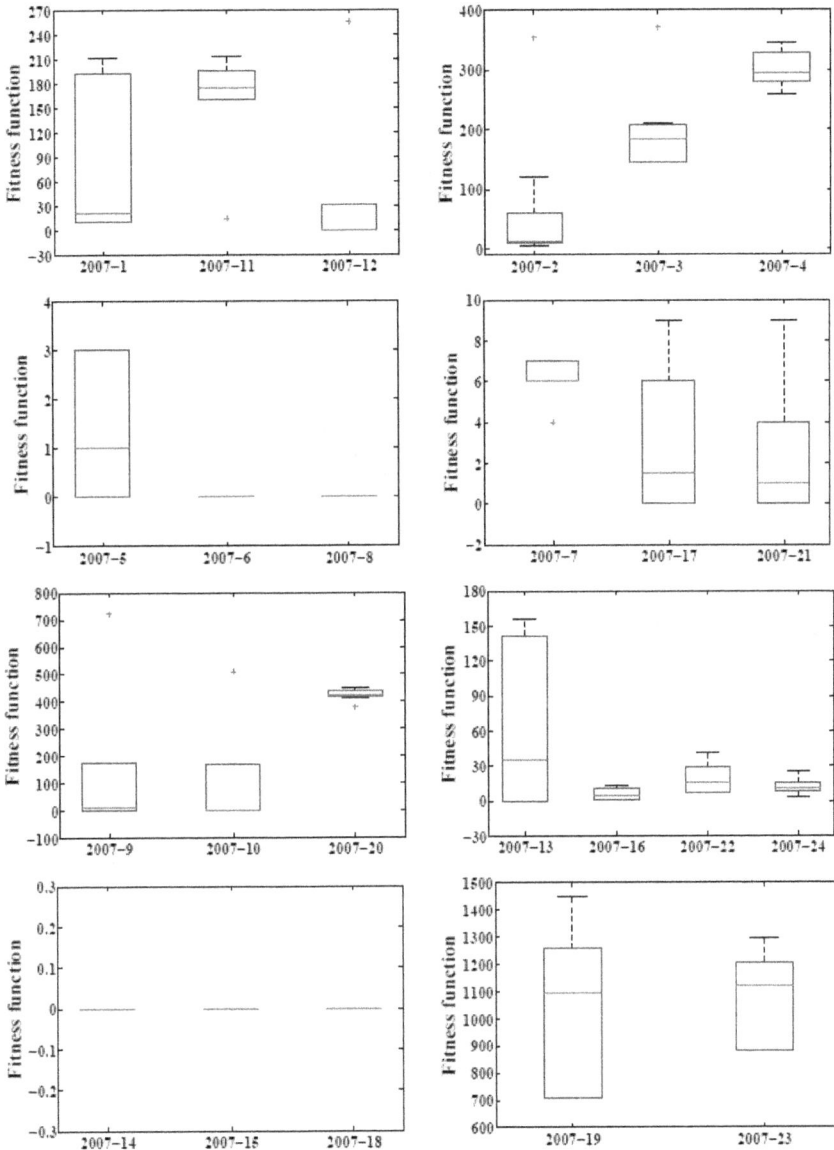

Figure 1.7 Boxplots of results obtained for ITC2007 problem instances.

From the obtained results, it can be easily observed that ACOWSG produces competitive results on both the considered sets of problem instances. Hence, it is concluded that the proposed algorithm using ACO along with student grouping can be a suitable choice for solving UCTP.

1.7 CONCLUSIONS

This work proposed an innovative hybridization scheme of the ACO approach with student groupings for the solution of the UCTP. The proposed method utilized the events taken by the group of students instead of the individual students to allocate them first to the timeslot and then to the appropriate room. We have conducted extensive experimental tests to assess the ACOWSG algorithm's effectiveness and then compared the results obtained with several other existing results. For this purpose, two different sets of standard problem instances are considered. In the first set of experiments performed over Socha's benchmark dataset of 11 problem instances, ACOWSG offered better results for all the problem instances than the general ACO algorithm. In comparing ACOWSG with 18 different approaches available in the literature, we obtained improved or similar quality outcomes for six instances. The ACOWSG algorithm's efficiency can be observed for the small-sized problem instances as it offered the optimal solution within 0.5 seconds for all of them.

In the second set of experiments, performed over the 24 problem instances of the second track of ITC2007, ACOWSG offered a feasible solution for all of them within the stipulated time. The proposed algorithm also provides improved or equal quality outcomes for 13 problem instances when compared with the 13 existing algorithms. The algorithm also offered the optimal solution for 10 out of these 13 problem instances. Hence, the experimental results validate that the proposed ACOWSG algorithm is significant in terms of efficiency and computational cost. In the future, we are exploring the possibility of utilizing the proposed algorithm for solving other variants of educational timetabling problems, such as school timetabling (ITC11) and examination timetabling (examination track of ITC07). The algorithm can also be explored for solving other variants of timetabling problems.

REFERENCES

[1]   Wren, A.: Scheduling, timetabling and rostering—a special relationship? In: *The Practice and Theory of Automated Timetabling*, pp. 46–75. Springer (1996)

[2]   Lewis, R.: A survey of metaheuristic-based techniques for university timetabling problems. *OR Spectrum* 30(1), 167–190 (2008)

[3]   Gotlieb, C.: The construction of class-teacher timetables. In: *Proceedings of the International Federation of Information Processing Congress*, vol. 62, pp. 73–77. IFIP (1962)

[4]   Bashab, A., Ibrahim, A.O., AbedElgabar, E.E., Ismail, M.A., Elsafi, A., Ahmed, A., Abraham, A.: A systematic mapping study on solving university timetabling problems using meta-heuristic algorithms. *Neural Computing and Applications* 32, 17397–174321 (2020)

[5]   Fonseca, G.H., Santos, H.G., Carrano, E.G., Stidsen, T.J.: Integer programming techniques for educational timetabling. *European Journal of Operational Research* 262(1), 28–39 (2017)

[6] Gülcü, A., Akkan, C.: Robust university course timetabling problem subject to single and multiple disruptions. *European Journal of Operational Research* **283**(2), 630–646 (2020)

[7] Méndez-Díaz, I., Zabala, P., Miranda-Bront, J.J.: An ILP based heuristic for a generalization of the post-enrollment course timetabling problem. *Computers & Operations Research* **76**, 195–207 (2016)

[8] Siddiqui, A.W., Raza, S.A.: A general ontological timetabling-model driven metaheuristics approach based on elite solutions. *Expert Systems with Applications* **170**, 114268 (2021)

[9] Song, T., Liu, S., Tang, X., Peng, X., Chen, M.: An iterated local search algorithm for the university course timetabling problem. *Applied Soft Computing* **68**, 597–608 (2018)

[10] Abdullah, S., Turabieh, H.: Generating university course timetable using genetic algorithms and local search. In: *Third International Conference on Convergence and Hybrid Information Technology (ICCIT'08)*, vol. 1, pp. 254–260. IEEE (2008)

[11] Assi, M., Halawi, B., Haraty, R.A.: Genetic algorithm analysis using the graph coloring method for solving the university timetable problem. *Procedia Computer Science* **126**, 899–906 (2018)

[12] Wijaya, T., Manurung, R.: Solving university timetabling as a constraint satisfaction problem with genetic algorithm. In: *Proceedings of the International Conference on Advanced Computer Science and Information Systems (ICACSIS 2009)*. Depok (2009)

[13] Atsuta, M., Nonobe, K., Ibaraki, T.: ITC–2007 Track2: An approach using general CSP solver. In: *Proceedings of the Practice and Theory of Automated Timetabling (PATAT 2008)*, pp. 19–22 (2008)

[14] Cambazard, H., Hebrard, E., O'Sullivan, B., Papadopoulos, A.: Local search and constraint programming for the post enrolment-based course timetabling problem. *Annals of Operations Research* **194**(1), 111–135 (2012)

[15] Junn, K.Y., Obit, J.H., Alfred, R.: A constraint programming approach to solving university course timetabling problem (UCTP). *Advanced Science Letters* **23**(11), 11023–11026 (2017)

[16] Burke, E.K., Mareček, J., Parkes, A.J., Rudová, H.: Decomposition, reformulation, and diving in university course timetabling. *Computers & Operations Research* **37**(3), 582–597 (2010)

[17] van den Broek, J., Hurkens, C.A.: An IP-based heuristic for the post enrolment course timetabling problem of the ITC2007. *Annals of Operations Research* **194**(1), 439–454 (2012)

[18] Glover, F.: Future paths for integer programming and links to artificial intelligence. *Computers & Operations Research* **13**(5), 533–549 (1986)

[19] Al-Betar, M.A., Khader, A.T.: A harmony search algorithm for university course timetabling. *Annals of Operations Research* **194**(1), 3–31 (2012)

[20] Awadallah, M.A., Bolaji, A.L., Al-Betar, M.A.: A hybrid artificial bee colony for a nurse rostering problem. *Applied Soft Computing* **35**, 726–739 (2015)

[21] Bolaji, A.L., Khader, A.T., Al-Betar, M.A., Awadallah, M.A.: University course timetabling using hybridized artificial bee colony with hill climbing optimizer. *Journal of Computational Science* **5**(5), 809–818 (2014)

[22] Chen, M., Tang, X., Song, T., Wu, C., Liu, S., Peng, X.: A tabu search algorithm with controlled randomization for constructing feasible university course timetables. *Computers & Operations Research* **123**, 105007 (2020)

[23] Goh, S.L., Kendall, G., Sabar, N.R.: Simulated annealing with improved reheating and learning for the post enrolment course timetabling problem. *Journal of the Operational Research Society* 70(6), 873–888 (2019)

[24] Jat, S.N., Yang, S.: A hybrid genetic algorithm and tabu search approach for post enrolment course timetabling. *Journal of Scheduling* 14(6), 617–637 (2011)

[25] Palembang, C.: Design of rescheduling of lecturing, using genetics-ant colony optimization algorithm. In: *IOP Conference Series: Materials Science and Engineering*, vol. 407, p. 012111. IOP Publishing (2018)

[26] Chiarandini, M., Birattari, M., Socha, K., Rossi-Doria, O.: An effective hybrid algorithm for university course timetabling. *Journal of Scheduling* 9(5), 403–432 (2006)

[27] Landa-Silva, D., Obit, J.H.: Great deluge with non-linear decay rate for solving course timetabling problems. In: *Fourth International Conference on Intelligent Systems (IS'08)*, vol. 1, pp. 8–11. IEEE (2008)

[28] Mcmullan, P.: An extended implementation of the great deluge algorithm for course timetabling. In: *International Conference on Computational Science (ICCS'07)*, pp. 538–545. Springer (2007)

[29] Abdullah, S., Burke, E.K., McCollum, B.: An investigation of variable neighbourhood search for university course timetabling. In: *Second Multidisciplinary International Conference on Scheduling: Theory and Applications (MISTA'05)*, pp. 413–427. MISTA (2005)

[30] Li, J.Q., Pan, Q.K., Wang, F.T.: A hybrid variable neighborhood search for solving the hybrid flow shop scheduling problem. *Applied Soft Computing* 24, 63–77 (2014)

[31] Lohpetch, D., Jaengchuea, S.: A hybrid multi-objective genetic algorithm with a new local search approach for solving the post enrolment-based course timetabling problem. In: *Recent Advances in Information and Communication Technology*, pp. 195–206. Springer (2016)

[32] Rezaeipanah, A., Abshirini, Z., Zade, M.B.: Solving university course timetabling problem using parallel genetic algorithm. *International Journal of Scientific Research in Computer Science and Engineering* 7(5), 5–13 (2019)

[33] Abdulkader, M.M., Gajpal, Y., ElMekkawy, T.Y.: Hybridized ant colony algorithm for the multi compartment vehicle routing problem. *Applied Soft Computing* 37, 196–203 (2015)

[34] Nothegger, C., Mayer, A., Chwatal, A., Raidl, G.R.: Solving the post enrolment course timetabling problem by ant colony optimization. *Annals of Operations Research* 194(1), 325–339 (2012)

[35] Al-Betar, M.A., Khader, A.T., Liao, I.Y.: A harmony search with multi-pitch adjusting rate for the university course timetabling. In: *Recent Advances in Harmony Search Algorithm*, pp. 147–161. Springer (2010)

[36] Malim, M.R., Khader, A.T., Mustafa, A.: Artificial immune algorithms for university timetabling. In: *Proceedings of the Sixth International Conference on Practice and Theory of Automated Timetabling*, pp. 234–245. Springer (2006)

[37] Dorigo, M., Maniezzo, V., Colorni, A.: Ant system: optimization by a colony of cooperating agents. *IEEE Transactions on Systems, Man, and Cybernetics, Part B (Cybernetics)* 26(1), 29–41 (1996)

[38] Socha, K., Knowles, J., Sampels, M.: A max-min ant system for the university course timetabling problem. In: *Ant Algorithms*, pp. 1–13. Springer (2002)

[39] Rossi-Doria, O., Sampels, M., Birattari, M., Chiarandini, M., Dorigo, M., et al.: A comparison of the performance of different metaheuristics on the timetabling problem. In: *Proceedings of the Fourth International Conference on Practice and Theory of Automated Timetabling*, pp. 329–351. Springer (2003)

[40] Ejaz, N., Javed, M.Y.: A hybrid approach for course scheduling inspired by die-hard co-operative ant behavior. In: *International Conference on Automation and Logistics*, pp. 3095–3100. IEEE (2007)

[41] Ayob, M., Jaradat, G.: Hybrid ant colony systems for course timetabling problems. In: *Second Conference on Data Mining and Optimization (DMO'09)*, pp. 120–126. IEEE (2009)

[42] Mazlan, M., Makhtar, M., Ahmad Khairi, A., Mohamed, M.A.: University course timetabling model using ant colony optimization algorithm approach. *Indonesian Journal of Electrical Engineering and Computer Science* **13**(1), 72–76 (2019)

[43] Lewis, R., Paechter, B.: Application of the grouping genetic algorithm to university course timetabling. In: *Evolutionary Computation in Combinatorial Optimization*, pp. 144–153. Springer (2005)

[44] Falkenauer, E.: *Genetic Algorithms and Grouping Problems*. John Wiley & Sons, Inc. (1998)

[45] Badoni, R.P., Gupta, D., Mishra, P.: A new hybrid algorithm for university course timetabling problem using events based on groupings of students. *Computers & Industrial Engineering* **78**, 12–25 (2014)

[46] Papadimitriou, C.H., Steiglitz, K.: *Combinatorial Optimization: Algorithms and Complexity*. Courier Dover Publications (1998)

[47] Bonabeau, E., Dorigo, M., Theraulaz, G.: *From Natural to Artificial Swarm Intelligence*. Oxford University Press, Inc. (1999)

[48] Abdullah, S., Burke, E.K., McCollum, B.: Using a randomised iterative improvement algorithm with composite neighbourhood structures for the university course timetabling problem. In: *Metaheuristics*, pp. 153–169. Springer (2007)

[49] Burke, E.K., McCollum, B., Meisels, A., Petrovic, S., Qu, R.: A graph-based hyper-heuristic for educational timetabling problems. *European Journal of Operational Research* **176**(1), 177–192 (2007)

[50] Abdullah, S., Burke, E.K., McCollum, B.: A hybrid evolutionary approach to the university course timetabling problem. In: *IEEE Congress on Evolutionary Computation (CEC'07)*, pp. 1764–1768. IEEE (2007)

[51] Asmuni, H., Burke, E.K., Garibaldi, J.M.: Fuzzy multiple heuristic ordering for course timetabling. In: *Proceedings of the Fifth United Kingdom Workshop on Computational Intelligence (UKCI'05)*, pp. 302–309. Citeseer (2005)

[52] Abdullah, S., Turabieh, H., McCollum, B., McMullan, P.: A hybrid metaheuristic approach to the university course timetabling problem. *Journal of Heuristics* **18**(1), 1–23 (2012)

[53] Abuhamdah, A., Ayob, M.: Average late acceptance randomized descent algorithm for solving course timetabling problems. In: *International Symposium in Information Technology (ITSim'10)*, vol. 2, pp. 748–753. IEEE (2010)

[54] Landa-Silva, D., Obit, J.H.: Evolutionary non-linear great deluge for university course timetabling. In: *Hybrid Artificial Intelligence Systems*, pp. 269–276. Springer (2009)

[55] Rezaeipanah, A., Matoori, S.S., Ahmadi, G.: A hybrid algorithm for the university course timetabling problem using the improved parallel genetic algorithm and local search. *Applied Intelligence* **51**(1), 467–492 (2021)

[56] Lewis, R.: A time-dependent metaheuristic algorithm for post enrolment-based course timetabling. *Annals of Operations Research* **194**(1), 273–289 (2012)

[57] Cambazard, H., Hebrard, E., O'Sullivan, B., Papadopoulos, A.: Local search and constraint programming for the post enrolment-based course timetabling problem. In: *Proceedings of the Seventh International Conference on the Practice and Theory of Automated Timetabling*. PATAT (2008)

[58] Chiarandini, M., Fawcett, C., Hoos, H.H.: A modular multiphase heuristic solver for post enrollment course timetabling. In: *Proceedings of the Seventh International Conference on the Practice and Theory of Automated Timetabling*. PATAT (2008)

[59] Müller, T.: ITC2007 solver description: A hybrid approach. *Annals of Operations Research* **172**(1), 429–446 (2009)

[60] Ceschia, S., Di Gaspero, L., Schaerf, A.: Design, engineering, and experimental analysis of a simulated annealing approach to the post-enrolment course timetabling problem. *Computers & Operations Research* **39**(7), 1615–1624 (2012)

[61] Soria-Alcaraz, J.A., Ochoa, G., Swan, J., Carpio, M., Puga, H., Burke, E.K.: Effective learning hyper-heuristics for the course timetabling problem. *European Journal of Operational Research* **238**(1), 77–86 (2014)

[62] Soria-Alcaraz, J.A., Ochoa, G., Sotelo-Figueroa, M.A., Carpio, M., Puga, H.: Iterated VND versus hyper-heuristics: Effective and general approaches to course timetabling. In: *Nature-Inspired Design of Hybrid Intelligent Systems*, pp. 687–700. Springer (2017)

Chapter 2

Genetic optimization and their applications

Tusar Kanti Mishra, Vaibhav Bhartia, and B. K. Tripathy

2.1 INTRODUCTION

By the word "optimum", we mean "maximum" or "minimum" depending upon the circumstances. During the optimization process, the best is obtained if it is possible to measure and change what is "good" or "bad". Optimization theory is the branch of mathematics encompassing the quantitative study of optima and the methods for finding them. Optimization brings in effectiveness and extent of goodness in problem-solving [1]. Optimization practice, on the other hand, is the collection of techniques, methods, procedures, and algorithms that can be used to find the optima. Optimization has been a prolonged challenging topic for researchers for centuries. Literature suggests that the foundation for optimization was laid long before the Christ era. The first-ever optimization technique aimed to calculate the closeness among a pair of points.

Presently, optimization has a significant existence as a robust tool in almost every field of applications. The omnipresence effect of optimization is well realized in engineering, medical science, agriculture, applied sciences, mathematics, data science, space computation, education management, construction, and many more. Typical areas of application are modeling, characterization, and design of devices, circuits, and systems; design of tools, instruments, and equipment; design of structures and buildings; process control, approximation theory, curve fitting, solutions of systems of linear equations, forecasting, production scheduling, quality control, maintenance and repair, inventory control, accounting, budgeting, etc. The justification behind the vivid utilization of optimization relies on the ease of use and assurance of guaranteed solutions with computationally efficient efforts. Researchers are finding it an eternal tool whereby the upbringing of novel methodologies is still on. Some recent innovations rely almost entirely on optimization theory, for example, neural networks and adaptive systems. The workflow diagram for the selection and use of optimization methods in any application is presented in Figure 2.1. In the beginning, the sub-problems/components needed to be optimized are identified. These sub-problems identify the design variables that need to be optimized. These

DOI: 10.1201/9781003462422-2

variables are subject to certain constraints and boundary values. The optimizations are to be carried out under these constraints and boundary values. The optimization methods selected are supposed to work under these constraints and the boundary values and optimization problems involved. The results of these optimization methods are assessed to find whether they are generating satisfactory results or not. If the results are not satisfactory, then the workflow may have several reverse operations, which are not shown in the workflow diagram as these decisions are application specific. The selection of optimization methods is made after this step. Finally, the best possible results are collected, which are the results of these optimization methods.

In real-life problems, multiple solutions often exist. Sometimes this number goes up to infinity. For optimization methods to be applicable, it is inherent that the problem in hand has several solutions so that the best solution can be identified depending upon some performance criterion.

Broadly, the optimization methods can be categorized into the following four categories:

1. Analytical methods
2. Graphical methods
3. Experimental methods
4. Numerical methods

Techniques of differential calculus are used in the analytical methods. Before applying the rules of calculus, the problem to be solved is needed to be described in mathematical terms. Digital computers are not necessary to

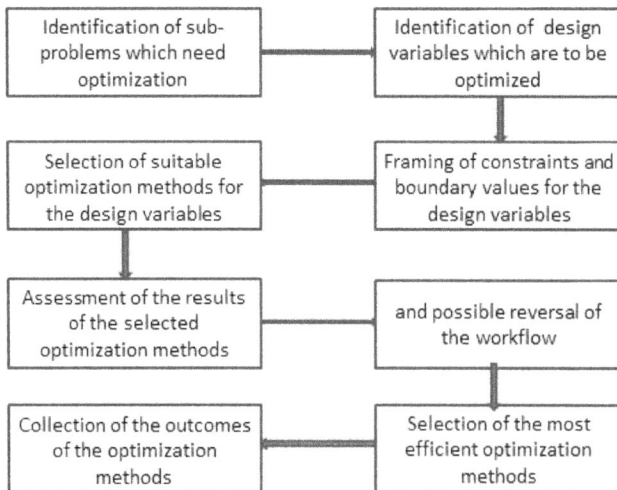

Figure 2.1 Major steps in the workflow of an optimization process.

solve these problems. We shall deal with these methods in detail in a subsequent section. However, these methods are not efficient enough to be applied to highly nonlinear problems and also to problems involving two or three independent parameters.

A graphical method can be used to plot the functions to be minimized or maximized if the number of variables does not exceed 2. If the function depends upon one variable, say "x", a plot of "x" versus $f(x)$ will immediately reveal the maxima and/or the minima of the function. A similar kind of technique is followed when the function depends upon two variables by constructing a set of contours. A contour is a set of points in the plane having specific properties. So a contour plot, like a topographical map of a specific region, will readily reveal the function's peaks and valleys. But, as the functions are to be optimized depending upon several parameters, this method is not applicable.

In experimental methods, the process variables are adjusted in random order, and the performance criterion is measured at every step by the way reaching an optimum or a near optimum. However, it can lead to unreliable results, as in certain systems, two or more variables interact with each other and must be adjusted simultaneously to yield the optimum performance criterion.

Starting with an initial solution, iterative numerical methods are adopted to improve the solutions progressively after each step in the iterations, which ends when a stopping condition/convergence criteria is/are satisfied. The stopping condition may be when no more input is available, and the convergence criteria may be that there is no significant improvement in the performance criteria. This approach is perhaps the most important general approach to optimization.

The advantages of the Numerical methods are that these methods can be applied to optimization problems where the other three methods are not applicable and that one can use digital computers to program these methods. As a result, these are the most efficient and popular methods used for solving optimization problems.

Another method that provides a classification of optimization techniques is presented in Figure 2.2.

2.1.1 Difficulties faced

These methods converge quickly to optimal solutions. But they are not applicable to cases where the objective functions are not differentiable or discontinuous problems. The convergence to an optimal solution depends on the chosen initial solution. Most algorithms tend to get stuck to a suboptimal solution. The solution space may have several optimal solutions, many being locally optimal and generating low objective function values as the problem variables are mostly nonlinear and there are complex interactions among them.

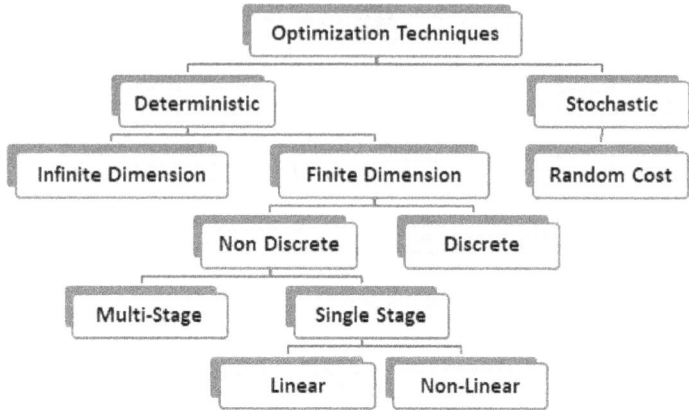

Figure 2.2 Classification of optimization techniques.

Moreover, finding the local optimal solutions is impossible once the process gets stuck up there. An algorithm efficient in solving one optimization problem may not be efficient in solving a different optimization problem. The traditional optimization algorithms are developed with a specific type of problem in view. One algorithm may be best suited for one problem, while it may not even apply to a different situation. It compels the designers to know more than one optimization method. Algorithms cannot be efficiently used on parallel machines. Because of the affordability and availability of parallel computing machines, it is now convenient to use parallel machines in solving complex engineering design optimization problems. However, since most of the traditional methods use a point-by-point approach, where one solution gets updated to a new solution in an iteration, the advantage of parallel machines cannot be exploited. The optimization methods assume that the problem variables are continuous, whereas, in reality, the variables are discrete only. So, to make the approaches applicable to these problems, it is assumed that the variables are continuous during the optimization process, and in the end, a value closer to the solution is recommended. The infeasible values of program variables are permitted; hence, the process has to go through all the infeasible solutions by spending a lot of time computing them. This increases the computational time, which is undesirable. The nearest upper and lower values for each of the program variables are to be computed at the end, which leads to the computation of a total number of such values. It comes to two options checked for each variable; hence, it may not guarantee the forming of the optimal combination with respect to other variables. All these difficulties can be eliminated if only feasible values of the variables are allowed during the optimization process.

2.1.2 Evolutionary algorithms

In order to find methods that do not have most of the above difficulties, biologically inspired algorithms (BIAs) have been found to be a suitable direction. BIAs mimic the intelligent behaviors of the biological behavior of animals, birds, fish, and so on. Bio-inspired algorithms are considered a major part of nature-inspired algorithms. They have become very important in a wide variety of applications in many domains. BIAs are motivated by the challenges in applications where conventional optimization techniques are not effective. There are three main branches of BIAs, namely evolutionary algorithms (EAs), swarm intelligence (SI), and bacterial foraging algorithms (BFAs), which are based on the understanding of biological systems, genetic evolution, animal behaviors, and bacterial foraging patterns. These algorithms have enjoyed great success in solving complex real-world problems in the evolution process; the selection of species is based on their ability to survive in an environment. Reproduction is also found to be accompanied by the mutation of species, which performs as one of the major components of evolution. Between the 1950s and 1960s, several evolutionary algorithms were proposed. Rochenberg and Schwefel introduced another novel optimization technique to solve aerospace engineering problems, later named an evolutionary strategy. Peter Bienert joined Rochenberg and Schwefel later to construct an automatic experimenter using simple rules of mutation and selection. There was no crossover in this technique. Only mutation was used to generate offspring, and an improved solution was kept at every generation. This was essentially a simple trajectory-style hill-climbing algorithm with randomization. In 1966, Fogel et al. developed an evolutionary programming technique by representing the solutions as finite-state machines and randomly mutating one of these machines.

The major types of evolutionary algorithms are:

- Genetic programming
- Evolutionary programming
- Evolutionary strategy
- Genetic algorithm

Lawrence J. Fogel in 1960 developed the concept of evolutionary programming (EP). The field of artificial intelligence revolves around the simulation of primitive neural networks and heuristics. As estimated by Fogel, these two approaches had limited scope as the focus was on modeling humans instead of focusing on the evolution of creatures of higher intellect. Adapting behavior was supposed to be the main characteristic of intelligence. The main ingredient of intelligent behavior was supposed to be the prediction. A series of experiments was conducted by simulating the evolution of finite-state machines. A series of other related experiments was conducted, and the documentation of these led to several publications between 1962 and 1968 [2].

Rochenberg and Schwefel introduced a novel optimization technique called evolutionary strategy to solve problems in aerospace engineering, Later, Peter Bienert joined Rochenberg and Schwefel later to construct an automatic experimenter using simple rules of mutation and selection. There was no crossover in this technique [3].

Genetic programming (GP) was derived from the model of biological evolution and is a new method for generating computer programs. A population of random programs is created initially, and the elements are continuously improved through a process of breeding. The new programs are generated through stochastic variations of the existing ones. The programs of GP systems evolve to solve pre-described automatic programming and machine learning problems.[4].

Genetic Algorithms are search and optimization procedures [5]. These are motivated by the principles of natural genetics and natural selection. Some fundamental ideas of genetics are borrowed and used artificially to construct search algorithms that are robust and require minimal problem information.

Its working principle is very different from that of most of the classical optimization techniques. It is assumed that all living creatures are descendants of older species. Any variation that occurs is due to natural selection. Some individuals have a greater chance of reproduction due to some heritable differences and are said to have higher "fitness". Fitness measures the success of an organism.

Exploring the scope of the usage of optimization has gained further momentum with the introduction of genetic algorithms. Genetic algorithm has been playing a major role since 1975. Study on genetic optimization techniques reveals the advancements in the front of computational efficiency for problem-solving, especially from the mathematical and computer science approaches. Our primary focus in this chapter is to present the optimization processes being carried out so far with the help of genetic algorithms.

2.2 CLASSICAL OPTIMIZATION TECHNIQUES

Classical optimization techniques are based on mathematical optimization models [6]. These models have been developed to describe the problem of minimizing or maximizing an objective function. These techniques are helpful in finding solutions to continuous and differentiable functions.

These techniques can be used for three types of problems:

- Single variable functions
- Functions with multiple variables having zero number of constraints
- Functions with multiple variables and having constraints involving the decision variables.

2.2.1 Single variable function optimization

Single-variable optimization can be defined as an optimization without constraints. In this optimization, there is only one decision variable for which we are trying to find a value in the domain of the function [7]. The general structure of such a problem is

$$\text{Min } f(x), x \in D \tag{2.1}$$

Here, $f(x)$ is called the objective function. The variable "x" is called the decision variable.

On many occasions, the optimization problem consists of finding the maximum of the objective function. Since the maximum of F can be readily obtained by finding the minimum of the negative of F and then changing the minimum sign. Consequently, in this and subsequent chapters, we focus on minimization without loss of generality.

2.2.2 Multi-variable function optimization

It is a generalization of the above case, where the number of decision variables is more than 1. Suppose that the decision variables are $x_1, x_2, ... x_n$. The objective function z can be written in the form [8].

$$z = f(x_1, x_2 .... x_n) \tag{2.2}$$

z may be linear or nonlinear. Hence, n variables can be manipulated or chosen to optimize this function z. It may be noted that one-dimensional optimization can be explained using pictures in two dimensions, since in the x-direction, we had the value of the decision variable, and in the y-direction, we had the value of the function. However, if it is a multidimensional optimization, it isn't easy to imagine in this manner.

In both the cases above, there were no restrictions on the decision variables. However, in real-life cases, we can have several constraints involving the decision variables.

2.2.2.1 Functions with multiple variables with constraints

No resource is infinite in character, and bounds exist for the decision variables involved in an objective function. Also, these variables are mostly related to each other. So, we have constraints that describe such relationships. In mathematics, equality is a relationship between two quantities or, more generally, two mathematical expressions, indicating that the quantities have the same value or that the expressions refer to the same mathematical object. Accordingly, there can be two types of constraints involved in an optimization problem: equality constraints and inequality constraints.

These optimization problems are called multivariate optimization with equality constraints and multivariate optimization with inequality constraints, respectively [9].

2.2.3 A general optimization algorithm

In this section, we present a general optimization algorithm. In the first step, the vector x_0 of the algorithm is initialized to a value estimated as the solution obtained by using knowledge about the problem at hand or initialized to "0" when no solution can be estimated. Till the convergence is achieved, Steps 2 and 3 are executed repeatedly in that order, called a single iteration. If the convergence is achieved, say after 'k' number of steps, the final step is executed to estimate the optimum solution to the problem. The vector $x^* = [x_1^*, x_2^* \ldots x_n^*]$ thus generated and the corresponding value of F, namely, $F^* = f(x^*)$ are output. The vector x^* is said to be the optimum, minimum, solution point, or simply the minimizer, and F^* is said to be the optimum or minimum value of the objective function.

The algorithm [10]

Step 1:
    (a) Set $k = 0$ and initialize x_0
    (b) Compute $F_0 = f(x_0)$
Step 2:
    (a) Set $k = k + 1$. Compute the changes in x_k by using an appropriate procedure given by the column vector Δx_k where $\Delta x_k^T = [\Delta x_1, \Delta x_2, \ldots \Delta x_n]$
    (b) Set $x_k = x_{k-1} + \Delta x_k$
    (c) Compute $F_k = f(x_k)$ and $\Delta F_k = F_{k-1} - F_k$
Step 3:
    Check if convergence has been achieved by using an appropriate criterion, for example, by checking ΔF_k and/or Δx_k. If this is the case, continue to step 4; otherwise go to step 2.
Step 4:
    (a) Output $x^* = x_k$ and $F^* = f(x^*)$
    (b) Stop

Figure 2.3 An optimization algorithm.

2.3 MODERN OPTIMIZATION TECHNIQUES

Several optimization methods have been developed that are conceptually different from conventional mathematical programming in recent years. These methods are called modern or non-traditional optimization methods. Most of these methods are based on specific properties and behaviors of biological, molecular, insect swarm, and neurological systems.

The different types of modern optimization techniques are:

- Simulated annealing (SA)
- Particle swarm optimization (PSO)
- Ant colony optimization (ACO)
- Honey bee optimization
- Genetic algorithms (GAs)

2.3.1 Simulated annealing

A random search technique for global optimization problems, simulated annealing (SA), demonstrates the annealing process in materials processing [11] as shown in Figure 2.3. This process causes the metal to cool and freeze into a crystalline state with minimum energy and larger crystallite size, reducing defects in the metal. The annealing process involves tight temperature control and a cooling rate called the annealing schedule [12].

Metaphorically, this is like dropping a few bouncing balls across a landscape, and as the balls bounce off and lose energy, they settle into some local minima. If a ball bounces often enough and loses energy slowly enough, some of the balls will eventually fall into the deepest part of the world. Therefore, it reaches the smallest in the world.

The basic idea of the simulated annealing algorithm is to use a Markov chain random search that not only allows changes that improve the objective function but also retains some changes that are not good [14].

In minimization problems, for example, all better moves or changes that reduce the value of the objective function are allowed. However, some changes (improvements) are also accepted with probability p.

Some of the characteristics of simulated annealing are:

- The standards of the results are unaffected by the starting estimates.
- Due to the discrete nature of functions and constraint evaluation, convergence or transition properties are not affected by the continuity or differentiability of the function.
- Convergence is also unaffected by the convexity of the feasible space
- Design variables need not be positive.

- This method can be used to solve mixed integer, discrete, or continuous problems.
- For problems with behavioral constraints (in addition to the lower and upper bounds of the design variables), as with genetic algorithms, an equivalent unconstrained function should be formulated.

2.3.2 Particle swarm optimization

Particle swarm optimization, abbreviated as PSO, is based on the colony or swarm behavior of insects such as ants, termites, bees, and wasps, flocks of birds; particle swarm optimization algorithms mimic the behavior of these social organisms [15]. The word particle means, for example, bees in a colony or birds in flock. Each individual or particle in the swarm operates in a distributed manner, using its intelligence and the collective or group intelligence of the swarm [16].

Therefore, once a particle finds a suitable route to food, the rest of the swarm can follow it immediately, even if they are far apart in the swarm. Swarm intelligence-based optimization techniques are called behavior-inspired algorithms [17]. Genetic algorithms, on the other hand, are called evolution-based procedures, which are initially done at random locations in a multidimensional design space. It is believed that every particle has two properties of her: position and velocity. Each particle wanders around the design space remembering the best position (in terms of a food source or objective function value) it found. The particles share information or preferred positions with each other and adjust their positions and velocities based on the preferred positions received.

2.3.3 Ant colony optimization

Ant is a socially addictive insect that adheres to organized colonies of individuals ranging from two million to 25 million. In [18], a metaheuristic method, which is a nature-inspired ant colony optimization (ACO), is proposed. A brief depiction of the flow process for the same is presented in Figure 2.4. When foraging, swarms of ants or mobile agents interact or communicate with their local environment. Each ant can deposit scent chemicals or pheromones to communicate with other ants, and each unit can also travel along roads marked with her pheromones laid by other ants [19]. When an ant finds a food source, it marks it with a pheromone and also marks the path to and from it. From the initial random foraging route, the pheromone concentration changes, and ants follow the route with higher pheromone concentration, and the pheromone is strengthened by increasing the number of ants. As more ants follow the same path, it becomes the preferred path [20]. This will create some favorite routes, often the shortest or more efficient. This is the positive feedback mechanism. Based on these features of ant behavior, a scientist has made significant progress in recent years, developing many powerful algorithms for ant colonies. Since then, many different variants have appeared. We can organize the category of new algorithms by

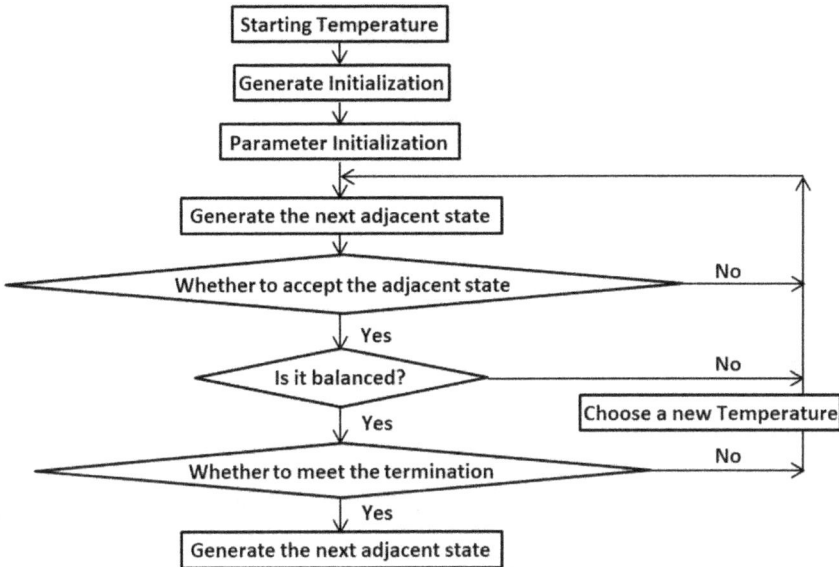

Figure 2.4 Typical overview of simulated annealing [13].

applying only some of the foraging behaviors of ants and adding some new features. ACOs are applied in routing in communication networks, salesman, graph mapping, scheduling, and constraint satisfaction problems.

2.3.4 Honey bee optimization

Bee algorithms form another class of algorithms closely related to ant colony optimization. The foraging behavior of the Bee inspires Bee's algorithm. Bees live in colonies, foraging honey and storing them in colonies they build [22]. A bee can communicate through pheromones and a" wiggle dance". For example, alarming bees may emit chemical messages (pheromones) to stimulate aggressive responses in other bees. Additionally, when a bee finds a good food source and brings nectar back to the hive, it communicates the location of the food source by performing a waving dance known as the signaling system. Such signal dances vary by species, but directional dances of varying strength are used to recruit more bees in an attempt to communicate the direction and distance of food resources found [23].

The bee algorithm assigns different food sources (or flower fields) to forage bees to maximize total nectar intake. Colonies should be optimized for the overall efficiency of nectar collection (Figure 2.5). Bee distribution depends on many factors, such as an abundance of nectar and proximity to hives. The probability of an observer bee chasing and foraging a dancing bee can be determined in different ways depending on the actual variant of the algorithm [25].

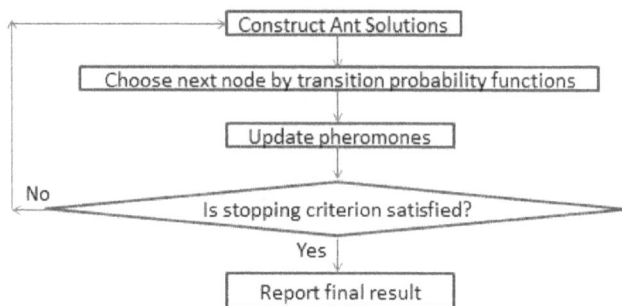

Figure 2.5 ACO flowchart in brief [21].

2.3.5 Genetic algorithms

Many of the practically optimal design problems are characterized by mixed continuous-discrete variables and discontinuous, non-convex design spaces [26]. Using standard nonlinear programming for this type of problem is inefficient and computationally expensive, and in most cases, it finds the closest relative optimum to the starting point [27]. Genetic algorithms (GAs) are well suited to solve such problems and, in most cases, can find the global optimum with high probability. Genetic Algorithms are search and optimization procedures. These are motivated by the principles of natural genetics and natural selection. Some fundamental ideas of genetics are borrowed and used artificially to construct robust search algorithms that require minimal problem information. A typical overview of GA is presented in Figure 2.7. The Darwinian principle of "Survival of the fittest" is the basic principle of GA. Also, it is taken that all living creatures are descendants of older species, and any variation that occurs is due to natural selection. Some individuals have a greater chance of reproduction due to some heritable differences and are said to have higher "fitness". Fitness measures the success of an organism.

The genetic mechanism behind evolution was put forth by Gregory Mendel in the 20th century. Each cell contains chromosomes, which are strands (twisted together) of DNA and consist of genes. Genes are the smallest hereditary unit. In biology, a gene is a sequence of DNA or RNA that codes for a molecule that has a function. Genes are inherited from parents. This inheritance happens during reproduction, and the process is called crossover. The set of all chromosomes is called the genome. In humans, genes vary in size from a few hundred DNA bases to more than two million bases.

As can be seen from Figure 2.7, there are three genetic operators [28]:

- Reproduction operator
- Crossover operator
- Mutation operator

Figure 2.6 Workflow for honey bee optimization [24].

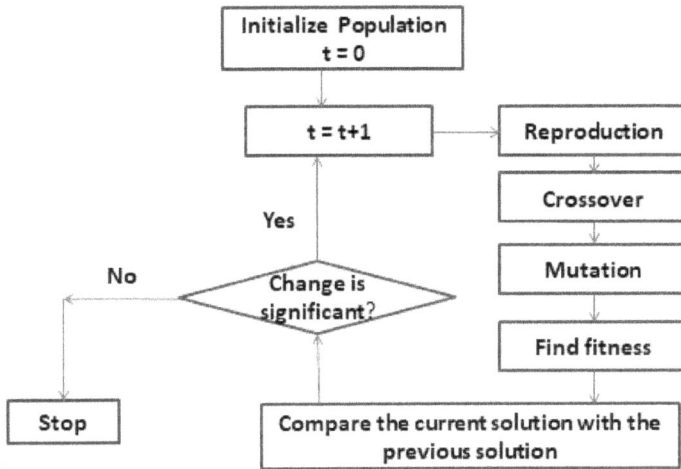

Figure 2.7 Typical overview of GA [29].

2.3.5.1 Reproduction operator

The primary objective of the reproduction operator is to emphasize reasonable solutions and eliminate bad solutions in a population while keeping the population size constant. Good (usually above average) solutions in a population are identified in this process. Multiple copies of good solutions are made, and these solutions replace bad solutions from the population. The reproduction operator is also called the selection operator. Actually, no new elements are generated in this process. It only puts more copies of good solutions at the expense of not-so-good solutions.

There are several reproduction/selection operators found in the literature. Some of these operators are:

- Roulette-wheel selection
- Boltzmann selection
- Tournament selection
- Rank selection
- Steady-state selection

2.3.5.2 Crossover operator

In almost all crossover operators, two strings are picked from the mating pool randomly, and some portions of the strings are exchanged between the strings. Indeed, every crossover between any two solutions from the new population is not likely to find children's solutions better than parent solutions. Still, the chance of creating better solutions is far better than random. If bad solutions are created, they get eliminated in the next reproduction operator and hence have a short life. In order to preserve some good strings selected during the reproduction operator, not all strings in the population are used in crossover. Suppose a crossover probability of p is used. In that case, 100p % strings in the population are used in the crossover operation, and $100(1 - p)\%$ of the population are simply copied to the new population. There exist a number of crossover operators in the GA literature:

- Single point crossover
- Two point crossover
- Multipoint crossover
- Uniform crossover, and so on

2.3.5.4 Mutation operator

The mutation operator changes a 1 to a 0 and vice versa with a small mutation probability, P_m.

The need for mutation is to keep diversity in the population. The solution obtained is better than the original solution. Although it may not happen all the time, mutating a string with a small probability is not a random operation since the process is biased for creating a few solutions in the neighborhood of the original solution.

GA differs from traditional optimization methods in the following ways:

- Instead of a single design point, a population of points (trial design vector) is used to start the course of action. If the number of design variables is n (usually), the population size is assumed to be between $2n$ and $4n$. More points than candidate solutions are used, so the GA is less likely to get stuck in a local optimum

- GA uses only objective function values. Derivatives are not used in the search step
- In GA, design variables are represented as strings of binary variables corresponding to chromosomes in natural genetics. Therefore, the search method is, of course, applicable to solving discrete and integer programming problems. With the continuous design variable, you can change the length of the string to achieve any resolution.
- Objective function values corresponding to design vectors play the role of fitness in natural genetics.
- For each new generation, a new string set is generated by a random selection of parents and crossover from the old generation (old string set).

2.4 PROPOSED METHOD

Neural networks are developed to model uncertainty in data and process them [30]. Adding more number of hidden layers to improve the accuracy and without increasing the complexity of computation, deep neural networks (DNNs) have been developed [31]. In the early years, there were concerns over determining an adequate number of hidden layers. But, later, Hinton in 2006 could provide solutions to this problem through his papers published in 2006. Since then, several different models of DNN have been proposed. These are convolutional neural networks (CNNs) [32], recurrent neural networks (RNNs), generative adversarial neural networks (GANs), and so on. Due to the advancement of several DNN models, there have been many improvements in research on artificial intelligence (AI). Many such applications have been presented in several works [33, 34]. There are many applications of deep learning like audio signal processing [35], image processing [36], COVID-19 research in the form of mask detection [37], classification [38], classification of skin cancer [39], text-based image retrieval [40], gene characterization [41], health care [42], brain MRI segmentation [43], computational biology [44], and detecting diabetic retinopathy [45]. Some general review papers describe several applications of deep learning [46, 47].

The proposed work emphasizes modifying the popular deep-learning CNN model [48]. In general, the CNN processes input through intermittent computations of convolution and pooling and finally carries out the classification using neural networks. This process sequence can be viewed as extracting the key feature points fixedly each time a pooling operation is performed, followed by convolution. It exhibits uniformity in selecting feature points. Although it ensures a good level of feature extraction, it does not ensure the most informative feature points extraction. The idea behind the proposed work focuses on this feat. The proposed work applies a mild alteration to the sequence of the process of CNN by introducing the genetic optimizer module in place of pooling. The inherent difference is that while pooling draws features at uniform behavior, the genetic optimizer

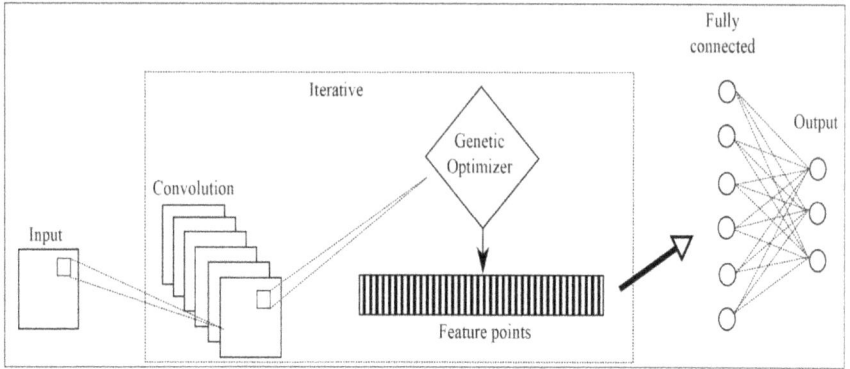

Figure 2.8 Block diagram of the proposed model (GO_CNN).

module compels the model always to extract the optimal feature points at any time. Thus, the final classification outcomes could always perform at a rate equivalent to or greater than conventional CNN. A block diagram of this explanation is depicted in Figure 2.8. Brief description of each process is presented later in a sequel.

2.4.1 Input

The input samples are basically standard images that demand classification. As a case study, facial expression images are considered in this work. However, the proposed mechanism can also be suitably utilized for other image samples. The facial expression images have been derived from the standard database (FER2013) which is derived from [49].

2.4.2 Convolution

Most often, the inputs to CNN are digital images [50]. Convolution, the first layer, performs the convolution operation on the input feeds. A specific size (say $d_1 \times d_2$) of filtering is applied to the sample. While the filter scrolls over the image, the dot products are computed and buffered into a matrix. Especially, prominent pixel level changes are obtained while this operation is carried out. This matrix is passed to the next phase, where the genetic optimizer works on it.

2.4.3 Genetic optimizer for CNN (GO$_{CNN}$)

The use of genetic optimization ensures the finest pixels (as obtained during pooling) and the nearby value finer pixels that can contribute toward the precision measure. The idea is to retain the pixel information that meets the fitness criteria. Say, talking about the fitness criteria in line with the average pooling process, one can idealize the retention of all those pixels whose values are not only exact to the local average but also nearby the local average of the current filtering window. Fitness, in this case, can be defined as:

$$\text{Fit}(\text{Ch}) = \text{Avg} - \delta \leq \text{intensity}(\text{Ch}) \leq \text{Avg} + \delta \tag{2.3}$$

subject to

$$0 \leq \text{Avg} - \delta \tag{2.4}$$

and

$$\text{Avg} - \delta \leq \text{Max} - \text{intensity} \tag{2.5}$$

where

Fit(Ch): Fitness evaluation mapping for chromosome Ch

Avg: Average intensity of the local convolution window resulting i-th in chromosome

δ: User specified deviation

Max −intensity: Highest intensity of the entire image.

Instead of applying the pooling operation, here genetic optimization is performed on the input matrix. The entry rows of the input matrix act as population samples for the optimizer. The pseudo-code for this is depicted in Algorithm 1.

Algorithm 1: $GO_{CNN}(M_p, d_1, d_2, R_c)$

Data: M_p : Input matrix,

d_1 : Number of rows acts as population size,

d_2 : Number of columns acts as chromosome length,

R_c : Rate of crossover.

Result: Mapped feature matrix M_f with optimal

features.

while *Constraint NOT satisfied* **do**

  for k_1 *in range 0 to* d_1 **do**

    for k_2 *in range 0 to* d_2 **do**

      Choose n number of chromosomes

      $\{ch_1, ch_2, \ldots, ch_n\}$

      Mutate $(ch_1, ch_2, \ldots, ch_n)$

      and obtain samples $(ch_1^1, ch_2^1, \ldots, ch_n^1)$

      Apply crossover

      and obtain samples $(ch_1^2, ch_2^2, \ldots, ch_n^2)$

    end

  end

  for p *in range 1 to* n **do**

    if ch_p^2 *is FIT* **then**

      $Append(M_f, ch_p^2)$;

    end

    if ch_p^2 *is !FIT* **then**

      $Discard(ch_p^2)$;

    else

  end

  $Return(M_f)$

end

The input matrix M_p is the intermediate result from the convolution layer of the CNN module. This M_p of dimension $d_1 \times d_2$ acts as the actual input to the GO$_{\text{CNN}}$. The rate of crossover parameter, that is, R_c, needs to be fed at the user's end. This parameter somehow relies on the length of the feature vector, hence providing the choice of value to the user for this is a good option. The algorithm works in accordance with the constraint as specified for a problem. This algorithm is generic in nature to be utilized for a vivid variety of scenarios. The expected output for this algorithm is to generate a mapped feature matrix M_f that contains the optimal feature points.

As depicted in this algorithm, the working steps start with the iteration directive to the constraint satisfaction. Among the available number of rows, certain rows are considered to be the pool of the population. Suitable genetic processes like mutation and crossover are applied to a population pool. The fitness evaluation according to the fitness function is implemented and the rows found under-fit are discarded. The complete process is thus implemented for each of the feeds of the convolution process applied on the input matrix. Simultaneously, the resultant fit number of rows forms the optimal feature matrix. The resultant matrix thus obtained is further subjected to a minor refinement phase, as explained later.

The refinement phase is necessarily implemented to comprehend the penalty of losing certain rows that might lead to the undersized matrix. To compensate, the nearby rank of rows is padded at appropriate positions until the matrix size regains its standard form. While evaluating the fitness, a ranking mechanism is also run in a spontaneous mode that takes care of the rank of each of the rows.

2.4.4 Case study on FER

A case study is carried out to validate the working of the proposed scheme. This is accomplished through pilot experimentation on set of facial expression images. For this purpose, the facial expression dataset (FER2013) is considered [49]. The benchmark dataset is referred from the FER2013 sample repository. Six distinct facial expression images are considered: angry, fear, happy, sad, surprised, and neutral. Certain samples are presented in Figure 2.9, where the labeling is done on integer ranges $[0, 1, \ldots, 5]$ for each expression class. The train and test split of the samples are considered to be 1,800 and 600, respectively, with a uniform number of samples from each category. Modeling the customized GO$_{\text{CNN}}$ is suitably carried out using the training samples. The parameters used for the purpose are presented as follows. M_p of dimension 64×64 is the actual input to the GO$_{\text{CNN}}$. The rate of crossover parameter, that is, R_c, needs to be kept as 6 (in this case, 10% of the row size) as in every iteration. The algorithm works in accordance with the constraint and fitness as specified in equations (1) and (2) earlier. As and when needed, the compensation for a certain number of rows is accomplished strictly with reference to the ranking mechanism, as stated earlier.

Figure 2.9 Sample snapshots of the facial expressions [49].

Table 2.1 Performance Indicators for CNN and GO$_{CNN}$, Respectively.

Fold#		CNN				GO$_{CNN}$			
	Size	T_P	F_P	T_N	F_N	T_P	F_P	T_N	F_N
1	30	12	3	12	3	13	2	12	3
2	30	13	2	13	2	14	1	13	2
3	30	12	3	13	2	12	3	13	2
4	30	14	1	14	1	14	1	13	2
5	30	12	3	12	3	13	2	12	3
6	30	13	2	13	2	13	2	13	2
7	30	15	0	14	1	14	1	14	1
8	30	14	1	15	0	14	1	15	0
9	30	13	2	13	2	13	2	14	1
10	30	14	1	13	2	14	1	13	2

The expected output feature matrix M_f is generated as per Algorithm 1. Finally, it is subjected to the classifier layer. The classifier layer utilizes a conventional backpropagation algorithm (BPNN) that classifies six distinct labels for the six categories of facial expression considered.

2.4.4.1 Experimental analysis

The same is validated, and the overall accuracy rate is computed following k-fold ($k = 10$) cross-validation. A comparison of the proposed GO$_{CNN}$ with the competent scheme (CNN) is carried out. Several performance indicators are considered as shown in Table 2.1. The ROC plots obtained from the resultants are presented in Figure 2.10. It can be observed that a mild increase in the AOC is noticed for both of the schemes under consideration.

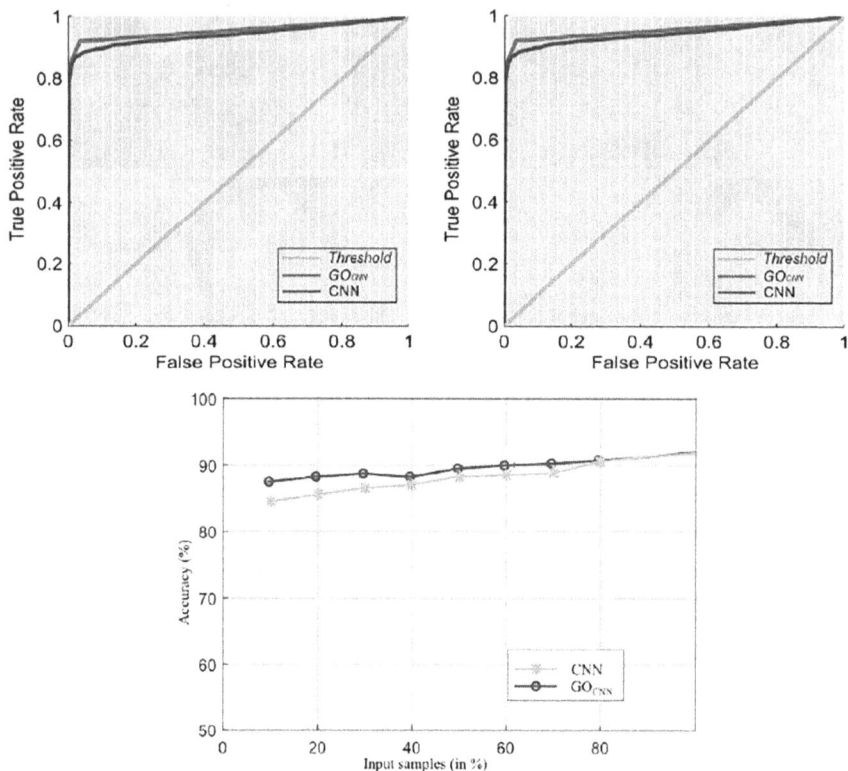

Figure 2.10 Comparison among the ROC and overall rate of accuracy for CNN and GO$_{CNN}$, respectively.

2.4.4.2 Discussion

Promising results in favor of the proposed work are obtained that outperform the competent scheme. In fact, when simulated on the said dataset, generic CNN yields an overall accuracy rate of 88%; however, the proposed model yields an increased accuracy rate of 88.5% on the same sample set. With this, suitable justification is built in favor of the proposed work. It strengthens the claim that genetic optimization has been a good choice where performance matters in terms of accuracy. The said work performs satisfactorily where the input sample image resolutions are very high (as nowadays applications on high-resolution images are drawing huge demands among researchers). The iterations of convolution to pooling (the case of CNN) are more when it comes to high-resolution images. However, the GO$_{CNN}$ model needs a slightly lesser number of iterations. For low-resolution image inputs, no difference in the computational time between CNN and GO$_{CNN}$ is observed.

2.5 CONCLUSION

In this chapter, milestones on various optimization methods are discussed with an emphasis on genetic optimization techniques. Furthermore, a novel customized scheme dubbed as GO CNN is proposed that signifies the use of genetic optimization in modern soft computing tools like CNN. A case study on a typical facial expression dataset is also briefly presented. The future scope is to focus on utilizing the proposed scheme on various datasets and prove forward the generic and consistent behavior of the same.

REFERENCES

[1] Adby, P.: *Introduction to optimization methods.* Springer Science Business Media (2013).

[2] Fogel, L. J., Owens, A. J. and Walsh, M. J.: *Artificial intelligence through simulated evolution.* Wiley (1966).

[3] Schwefel, H. P.: *Numerical optimization of computer models.* Wiley (1981).

[4] Corne, D. W. and Lones, M. A.: Evolutionary algorithms. In: *Handbook of heuristics.* Eds: R. Martí, P. Pardalos and M. Resende. Springer (2018).

[5] Banzhaf, W.: Artificial intelligence: Genetic programming. *International Encyclopaedia of the Social & Behavioural Sciences* (2001): 789–792.

[6] Persson, J. A., Paul, D., Stefan, J. J. and Fredrik, W. S.: *Combining agent based approaches and classical optimization techniques.* Third European Workshop on Multi-Agent Systems (2005).

[7] Pelikan, M.: *Bayesian optimization algorithm: From single level to hierarchy.* University of Illinois (2002).

[8] Rizzi, P.: *Optimization of multi-constrained structures based on optimality criteria.* 17th Structures, Structural Dynamics, and Materials Conference (1976), p. 1547.

[9] Polak, E. and David, M.: An algorithm for optimization problems with functional inequality constraints. *IEEE Transactions on Automatic Control* 21, no. 2 (1976): 184–193.

[10] Sinha, G. R.: *Modern optimization methods for science, engineering and technology: Introduction and background to optimization theory.* IOP Publishing (2019), pp. 1–18, 978-0-7503-2404-5.

[11] Daniel, D., Chaimatanan, S. and Mongeau, M.: Simulated annealing: From basics to applications. In: *Handbook of meta-heuristics.* Springer (2019), pp. 1–35.

[12] Bertsimas, D. and Tsitsiklis, J: Simulated annealing. *Statistical Science* 8, no. 1 (1993): 10–15.

[13] Tu, C., Liu, Y. and Zheng, L.: Hybrid element heuristic algorithm optimizing, neural network-based educational courses. *Wireless Communications and Mobile Computing* (2021): 1–12. https://doi.org/10.1155/2021/9581793.

[14] van Laarhoven, P. J. M. and Aarts, E. H. L.: Simulated annealing. In: *Simulated annealing: Theory and applications, mathematics and its applications,* vol. 37 (1987) Springer. https://doi.org/10.1007/978-94-015-7744-1_2.

[15] Wang, D., Tan, D. and Liu, L.: Particle swarm optimization algorithm: An overview. *Soft Computing* 22, no. 2 (2018): 387–408.

[16] Poli, R., Kennedy, J. and Blackwell, T.: Particle swarm optimization. *Swarm Intelligence* 1, no. 1 (2007): 33–57.

[17] Kennedy, J. and Russell, E.: Particle swarm optimization. In: *Proceedings of ICNN'95-international conference on neural networks*, vol. 4, IEEE (1995), pp. 1942–1948.

[18] Dorigo, M. and Thomas, S.: Ant colony optimization: Overview and recent advances. *Handbook of Metaheuristic* (2019): 311–351.

[19] Dorigo, M., Birattari, M. and Stutzle, T.: Ant colony optimization. *IEEE Computational Intelligence Magazine* 1, no. 4 (2006): 28–39.

[20] Blum, C.: Ant colony optimization: Introduction and recent trends. *Physics of Life Reviews* 2, no. 4 (2005): 353–373.

[21] Mohammadnia, A., Rahmani, R., Mohammadnia, S. and Bekravi, M.: A load balancing routing mechanism based on ant colony optimization algorithm for vehicular adhoc network. *International Journal Network and Computer Engineering* 8 (2016): 1–10.

[22] Dervis, K.:. *An idea based on honey bee swarm for numerical optimization*, vol. 200, Technical Report-tr06, Erciyes University, Engineering Faculty, Computer Engineering Department (2005).

[23] Fathian, M., Amiri, B. and Maroosi, A.: Application of honey-bee mating optimization algorithm on clustering. *Applied Mathematics and Computation* 190, no. 2 (2007): 1502–1513.

[24] Pham, D., Castellani, M. and Le Thi, H. A.: Nature-inspired intelligent optimisation using the bees algorithm. *Lecture Notes in Computer Science* 8342 (2014): 38–69. https://doi.org/10.1007/978-3-642-54455-22.

[25] Niazkar, M. and Afzali, S. H.: Closure to "assessment of modified honey bee mating optimization for parameter estimation of nonlinear Muskingum models" by Majid Niazkar and Seied Hosein Afzali". *Journal of Hydrologic Engineering* 23, no. 4 (2018): 07018003.

[26] Mirjalili, S.: *Genetic algorithm, evolutionary algorithms and neural networks.* Springer (2019), pp. 43–55.

[27] Kumar, M., Husain, D., Upreti, N. and Gupta, D.: Genetic algorithm: Review and application. *Journal of Information & Knowledge Management* (2010): SSRN 3529843.

[28] Sourabh, K., Chauhan, S. S. and Kumar, V: A review on genetic algorithm: Past, present, and future. *Multimedia Tools and Applications* 80, no. 5 (2021): 8091–8126.

[29] Papakostas, G. A., Koulouriotis, D. E., Polydoros, A. S. and Tourassis, V. D.: Evolutionary feature subset selection for pattern recognition applications. In: *Evolutionary algorithms.* InTech (2011). https://doi.org/10.5772/15655.

[30] Tripathy, B. K. and Anuradha, J.: *Soft computing—advances and applications.* Cengage Learning Publishers (2015). ASIN: 8131526194, ISBN-10:9788131526194.

[31] Bhattacharyya, S., Snasel, V., Hassanian, A. E., Saha, S. and Tripathy, B. K.: *Deep learning research with engineering applications.* De Gruyter Publications (2020). ISBN: 3110670909, 9783110670905. https://doi.org/10.1515/9783110670905.

[32] Maheswari, K., Shaha, A., Arya, D., Tripathy, B. K. and Raj Kumar, R.: Convolutional neural networks: A bottom-up approach. In: *Deep learning research*

with engineering applications. Eds: S. Bhattacharyya, A. E. Hassanian, S. Saha and B. K. Tripathy. De Gruyter Publications (2020), pp. 21–50. https://doi.org/10.1515/9783110670905-002.

[33] Adate, A., Tripathy, B. K., Arya, D. and Shaha, A.: *Impact of deep neural learning on artificial intelligence research*. In: *Deep learning research and applications*. Eds: S. Bhattacharyya, A. E. Hassanian, S. Saha and B. K. Tripathy. De Gruyter Publications (2020), pp. 69–84. https://doi.org/10.1515/9783110670905-004.

[34] Tripathy, B. K. and Adate, A.: *Impact of deep neural learning on artificial intelligence research*. Eds: D. P. Acharjya et al., Chapter 8. Springer Publications (2021).

[35] Bose, A. and Tripathy, B. K: Deep learning for audio signal classification. In: *Deep learning research and applications*. Eds: S. Bhattacharyya, A. E. Hassanian, S. Saha and B. K. Tripathy. De Gruyter Publications (2020), pp. 105–136. https://doi.org/10.1515/9783110670905-00660.

[36] Adate, A. and Tripathy, B. K.: Deep learning techniques for image processing. In: *Machine learning for big data analysis*. Eds: S. Bhattacharyya, H. Bhaumik, A. Mukherjee and S. De. De Gruyter Publications (2018), pp. 69–90.

[37] Yagna, S. S., Geetha Rani, T. K. and Tripathy, B. K.: Social distance monitoring and face mask detection using deep learning. In: *Computational intelligence in data mining, smart innovation, systems and technologies*. Eds: J. Nayak, H. Behera, B. Naik, S. Vimal and D. Pelusi, vol. 281. Springer (2022). https://doi.org/10.1007/978-981-16-9447-9_36.

[38] Sihare, P., Khan, A. U., Bardhan, P. and Tripathy, B. K: COVID-19 detection using deep learning: A comparative study of segmentation algorithms. In: *Proceedings of the 4th international conference on computational intelligence in pattern recognition (CIPR)*. Eds: A. K. Das et al. CIPR, LNNS 480 (2022), pp. 1–10.

[39] Jain, S., Singhania, U., Tripathy, B. K., Nasr, E. A., Aboudaif, M. K. and Ali, K. K.: Deep learning based transfer learning for classification of skin cancer. *Sensors* 21, no. 23 (December 6, 2021): 8142.

[40] Singhania, U. and Tripathy, B. K.: *Text-based image retrieval using deep learning*, Fifth Edition. Encyclopaedia of Information Science and Technology (2021), p. 11. https://doi.org/10.4018/978-1-7998-3479-3.ch007.

[41] Gupta, P., Bhachawat, S., Dhyani, K. and Tripathy, B. K.: A study of gene characteristics and their applications using deep learning. In: *Studies in big data*. Eds: S. Sekhar Roy and Y.-H. Taguchi, vol. 103, Chapter 4. Handbook of Machine Learning Applications for Genomics (2021). ISBN: 978-981-16-9157-7, 496166_1_En.

[42] Kaul, D., Raju, H. and Tripathy, B. K.: Deep learning in healthcare. In: *Deep learning in data analytics: Recent techniques, practices and applications, studies in big data*. Eds: D. P. Acharjya, A. Mitra and N. Zaman, vol. 91. Springer (2022), pp. 97–115. https://doi.org/10.1007/978-3-030-75855-4_6.

[43] Tripathy, B. K., Parikh, S., Ajay, P. and Magapu, C.: Brain MRI segmentation techniques based on CNN and its variants. In: *Brain tumour MRI image segmentation using deep learning techniques*. Ed: J. Chaki, Chapter 10. Elsevier Publications (2022), pp. 161–182. https://doi.org/10.1016/B978-0-323-91171-9.00001-6.

[44] Bhardwaj, P., Guhan, T. and Tripathy, B. K.: Computational Biology in the lens of CNN. I: *Studies in big data*. Eds: S. Sekhar Roy and Y.-H. Taguchi, vol.

103, Chapter 5. Handbook of Machine Learning Applications for Genomics (2021). ISBN: 978-981-16-9157-7 496166_1_En.

[45] Prabhavathy, P., Tripathy, B. K. and Venkatesan, M.: Analysis of diabetic retinopathy detection techniques using CNN models. In: *Augmented intelligence in healthcare: A pragmatic and integrated analysis. Studies in computational intelligence*. Eds: S. Mishra, H. K. Tripathy, P. Mallick and K. Shaalan, vol. 1024. Springer (2022). https://doi.org/10.1007/978-981-19-1076-0_6.

[46] Rungta, R. K., Jaiswal, P. and Tripathy, B. K.: A deep learning based approach to measure confidence for virtual interviews. In: *Proceedings of the 4th international conference on computational intelligence in pattern recognition (CIPR)*. Eds: A. K. Das et al. CIPR, LNNS 480 (2022), pp. 278–291.

[47] Adate, A. and Tripathy, B. K.: A survey on deep learning methodologies of recent applications. In: *Deep learning in data analytics – recent techniques, practices and applications), studies in big data*. Eds: D. P. Acharjya, A. Mitra and N. Zaman, vol. 91. Springer (2022), pp. 145–170. https://doi.org/10.1007/978-3-030-75855-4_9.

[48] Albawi, S., Mohammed, T. A. and Al-Zawi, S.: Understanding of a convolutional neural network. In: *2017 international conference on engineering and technology (ICET)*. IEEE (2017), pp. 1–6.

[49] Zahara, L., Musa, P., Prasetyo, W. E., Karim, I. and Musa, S. B.: *The facial emotion recognition (FER-2013) dataset for prediction system of micro-expressions face using the convolutional neural network (CNN) algorithm based raspberry Pi*. 2020 Fifth International Conference on Informatics and Computing (ICIC) (2020), pp. 1–9. https://doi.org/10.1109/ICIC50835.2020.9288560.

[50] Keresztes, P., Zarandy, A., Roska, T., Szolgay, P., Bezak, T., Hidvegi, T., Jonas, P. and Katona, A.: An emulated digital CNN implementation. *Journal of VLSI Signal Processing Systems for Signal, Image and Video Technology* 23, no. 2 (1999): 291–303.

Chapter 3

Different variants of unreliable server

An economic approach

Shreekant Varshney, Chandra Shekhar,
Vivek Tiwari, and Kocherlakota Satya Pritam

3.1 INTRODUCTION

Queues (or waiting lines) help amenities and industries/trades by providing orderly service. Forming a waiting line is a social phenomenon; it is helpful to society if it can be adequately organized so that both the unit/person that waits and the one that serves get the maximum benefit. For instance, there was a time when in airline terminals, passengers formed separate waiting lines in front of check-in counters. But now, we consistently see only one line feeding into several counters. This results from the understanding that a single-line strategy better serves travelers and the airline administration. Such an inference has come from examining how a waiting line is formed and the service is provided. The investigation is based on building a mathematical model representing the arrival process of persons/travelers who join the waiting line, the rules by which they are permitted into service, and the time it takes to serve the persons/travelers. Queueing theoretic approach represents the entire scope of models covering all apparent systems that incorporate characteristics of a waiting line. We categorize the demanding unit service, whether it is human or anything, as a customer. The unit providing service is known as the server. This nomenclature of customers and servers is used broadly, irrespective of the nature of the physical context.

Over the past few years, queueing-based service systems with server breakdowns have been an exciting topic for queueing theorists, decision-makers, policyholders, researchers, and practitioners. These service systems are extensively used in many realistic waiting line models in our day-to-day life, such as in computer and communication systems, production systems, supply chain management, inventory control, transportation systems, and flexible manufacturing systems. Many researchers/policymakers have published research papers on queueing-based service systems with server breakdowns over the past few years. Gaver [1] and Shogan [2] explored many waiting line models with server breakdowns and implemented the closed form/explicit expressions of numerous system performance

DOI: 10.1201/9781003462422-3

measures. Jayaraman et al. [3] considered a bulk queueing model with server breakdown and state-dependent arrival rate of incoming customers. In 2004, Ke [4] deliberated a control policy for the batch arrival queue with a primary start-up and server breakdown. Later, Jain and Agrawal [5] examined a single-server batch arrival queueing model with multiple states of the unreliable server under N-policy. In the same year, Wang and Yang [6] studied a controllable queueing system with an unreliable server. They demonstrated many motivating numerical investigations for optimal control and economic analysis using the semi-classical optimizer, the Quasi-Newton method. For more in-depth investigation, one can refer to the research findings (cf. [7, 8, 9, 10, 11, 12, 13, 14, 15]) and references therein. Afterward, Ke et al. [16] envisaged a feedback retrial queueing system with the impatience behavior of the customer & unreliable server. They developed a cost optimization problem by defining the optimal parameter situation under the stability condition using the probability global search Lausanne (PGSL) method. A retrial queueing model with a finite number of sources and customer collision is analyzed by Nazarov et al. [17]. They proved that the transient probability distribution of the number of customers follows a Gaussian distribution.

A new class of queueing models with the working breakdown of the server was introduced by Kalidas and Kasturi [18] in 2012. In the active breakdown state, the server can be broken down at any instant when it is in function. However, instead of stopping the service altogether, the server continues to provide the service at a lesser service rate. Liu and Song [19] modified the work of [18] to batch arrival queue and derived the probability generating function (PGF) of the stationary queue length and its stochastic decomposition. Again, Yang [20] extended the work in [18], combining server vacations and using the Runge–Kutta method of fourth order (RK-4 method) to calculate the time-dependent state probability distribution. A similar type of research was investigated by Purohit et al. [21] and Sharma and Gireesh [22] to introduce transient-state probability distribution and several system performance measures. Later, several research papers on performance characteristics of the queue-based service/ machining systems with the unreliable server using some other queueing nomenclatures have been reviewed by many authors (cf. [23], [24], [25], [26], [27]).

During the busy period of the server, some impatient customers may depart from the service system due to their high level of impatience when they make the service attempts and then meet all servers break down or break down. The aforementioned queueing phenomenon is known as customer abandonment (reneging). For some critical results and findings based on different variants of impatience customers, interested researchers can refer to Chang et al. [14], Alnowibet et al. [28], Danilyuk et al. [29], Chai et al. [30], and Ehsan et al. [31].

Motivated by the above situations, the developed model 3 in this chapter analyzes a queueing scenario with customer impatience, the working breakdown of the server with service pressure conditions, and the threshold-recovery policy. The concept of threshold-based recovery policy was first introduced by Efrosinin and Semenova [32]. In threshold-based recovery policy, the server is broken down unpredictably while providing the service for the customer. Furthermore, repairing the server cannot be initiated until a threshold number of customers are present in the service system. In addition, Effrosinin and Winkler [33] investigated the single-server retrial queueing system with an unreliable server and threshold-based recovery policy. For a more detailed analysis, readers can refer to [21], [34], [35], [36], [37], and the research articles mentioned therein.

Nowadays, the demand for quality service is difficult for an entrepreneur to control. Due to this, the customers have to wait in the queue till sufficient service capability is available. Therefore, increasing the service quality in the quick-service industry is the most appropriate way to keep the enterprise's competitiveness. Furthermore, continuously serving customers is the primary job of the forefront attendants; whenever too many customers wait for their service, the attendants' working pressure increases. As a result, the generated pressure will influence the service rate.

Meanwhile, to minimize the excess of waiting customers, we consider a situation where the service rate increases when more customers wait for their service in the service system. And a service pressure coefficient models this feature appropriately for the service systems. The service pressure coefficient is a positive constant value indicating how the servers increase the service rate to reduce the number of waiting customers quickly. To analyze and investigate the service systems with service pressure coefficient, several studies in Markovian and non-Markovian environments have been done by many researchers and scientists (cf. [38], [39], [40]).

In the forthcoming sections, we describe the effect of working breakdown, service pressure environment, and threshold-based recovery policy on some system characteristics. First, we classify our study into three parts: (i) service system with server breakdown, (ii) service system with the working breakdown of the server, and (iii) service system with working breakdown, service pressure environment, and threshold-based recovery policy. Then, we demonstrate the set of Chapman–Kolmogorov differential-difference equations and corresponding matrix representation for analysis purposes. Next, to compute the steady-state probability distribution, we employ the matrix approach as in Section 3.6.1. Moreover, to better understand the research findings from an economic perspective, the explicit closed-form expressions of several system performance indicators and the cost optimization problem (see Section 3.6.2) are formulated. In Section 3.7, several numerical experiments are illustrated, and research findings are depicted with the help of various graphs and tables. Finally, the concluding remarks and future perspectives are provided in Section 3.8, which help the system analysts, decision-makers, researchers, and queueing theorists to design the upgraded service and machining systems.

3.2 SERVICE SYSTEM WITH SERVER BREAKDOWN

Notations

- λ : Mean arrival rate of customers in the service system
- μ_b : Mean service rate during the busy period of the server
- α : Breakdown rate of the server
- β : Repair rate when the server is broken down
- K : Capacity of the service system

Over the past few decades, queueing-based service systems with breakdown servers have attracted queueing theorists, system designers, and researchers. It considers not only the service system's queueing characteristics but also the server's reliability indices. This notion is extensively applied in telecommunications, production, inventory, manufacturing, and computer systems. Readers may refer to Krishnamoorthy et al. [41] and references therein for a recent survey. The queueing literature with server breakdown assumes that the service stops entirely if the server is broken down. Therefore, inspired by this, we will develop a finite-capacity service system with an unreliable server. In the subsequent section, we perform a comparative analysis based on different regimes of unreliable servers and the expected total cost of the service system.

Next, we describe some basic assumptions for the single-server finite-capacity service system with server breakdown, which are as follows.

- Customers arrive in the service system following the Poisson process with parameter λ.
- Customers form a waiting queue based on their arrival to join the service system, that is, the *First Come First Serve* (FCFS) service discipline is adopted.
- The service times during the regular working attribute of the server is exponentially distributed with mean $\dfrac{1}{\mu_b}$.
- The active server can serve only one customer at a time. If the incoming customer finds the server busy, it has to wait until the server is available.
- It is assumed that the server can break down only when at least one customer is present in the service system. The breakdown times follow an exponential distribution with parameter α.
- When the server breaks down, the server is immediately repaired, and the repair times follow an exponential distribution with parameter β.

- The symbol K represents the capacity of the service system, and it is assumed that the capacity of the service system is finite, that is, $K < \infty$.
- All the stochastic processes involved in the service system are assumed to be independent.

This chapter presents the queueing analysis of a finite-capacity service system using different variants of unreliable servers. For that purpose, we define the states of the governing model at time instant t using the fundamental law of the Markov process as

$$J(t) = \begin{cases} 0; & \text{The server is in regular working mode at time instant } t \\ 1; & \text{The server is in a breakdown state at time instant } t \end{cases}$$

and

$$N(t) \equiv \text{Number of customers in the queue based}$$

$$\text{service system at time instant } t$$

Therefore, $X(t) = \{(J(t), N(t)); t \geq 0\}$ represents the continuous-time Markov chain (CTMC) on state space

$$\Theta_1 \equiv \{(j,n); j = 0,1 \ \& \ n = 1,2,\cdots,K\} \cup \{(0,0)\}$$

where K represents the capacity of the queue-based service system.

Therefore, the Markov chain $\{X(t); t \geq 0\}$ is irreducible. Since the state space Θ_1 is finite, the Markov chain is positive recurrent. To get a view of transitions between the precedence states of the service system, the state transition diagram for a basic finite capacity queueing model with server breakdown is provided in Figure 3.1.

Now, using the elementary concept of the quasi-birth and death (QBD) process and balancing the input and output flow rates in Figure 3.1, the set of governing Chapman–Kolmogorov differential-difference equations of the studied model is developed as follows:

- When there is no customer in the service system, and the server is idle

$$\frac{dP_{0,0}(t)}{dt} = -\lambda P_{0,0}(t) + \mu_b P_{0,1}(t) \tag{3.1}$$

- When there are n customers in the service system, and the server is busy in regular working mode

$$\frac{dP_{0,n}(t)}{dt} = -(\lambda + \mu_b + \alpha) P_{0,n}(t) + \lambda P_{0,n-1}(t)$$
$$+ \mu_b P_{0,n+1}(t) + \beta P_{1,n}(t); 1 \leq n \leq K-1 \tag{3.2}$$

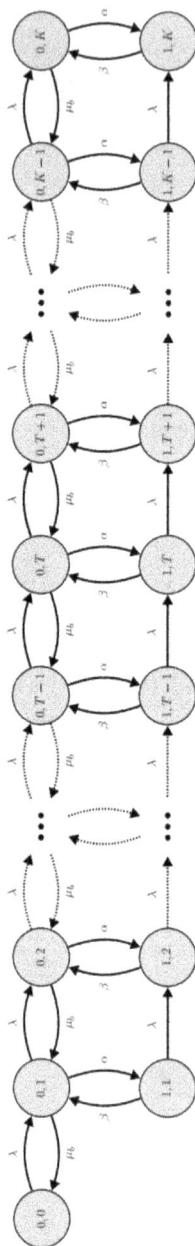

Figure 3.1 State transition diagram of a finite-capacity service system with server breakdown.

- When the capacity of the system becomes full, and the server is busy in regular working mode

$$\frac{dP_{0,K}(t)}{dt} = -(\mu_b + \alpha)P_{0,K}(t) + \lambda P_{0,K-1}(t) + \beta P_{1,K}(t) \tag{3.3}$$

- When there is only one customer in the service system, and the server is in the breakdown state

$$\frac{dP_{1,1}(t)}{dt} = -(\lambda + \beta)P_{1,1}(t) + \alpha P_{0,1}(t) \tag{3.4}$$

- When there are n customers in the service system, and the server is in the breakdown state

$$\frac{dP_{1,n}(t)}{dt} = -(\lambda + \beta)P_{1,n}(t) + \lambda P_{1,n-1}(t)$$
$$+ \alpha P_{0,n}(t); \quad 2 \leq n \leq K-1 \tag{3.5}$$

- When the capacity of the system becomes full, and the server is in the breakdown state

$$\frac{dP_{1,K}(t)}{dt} = -\beta P_{1,K}(t) + \lambda P_{1,K-1}(t) + \alpha P_{0,K}(t) \tag{3.6}$$

Furthermore, the matrix method is applied to determine the state probability distribution. For that purpose, the system of differential-difference equations (1)–(6) is transformed into the matrix form $DY = Q_1Y$, wherein Y is a column vector of all time-dependent probabilities of dimension $2K+1$, and DY is the derivative of the column vector Y. Similarly, Q_1 is the block-square transition rate matrix of order $2K+1$, which is generated by using the lexicographic characteristics of matrices. The diagonal-block structure of the transition rate matrix Q_1 is demonstrated as

$$Q_1 = \begin{bmatrix} A_{00}^1 & B_{01}^1 & 0 & \cdots & 0 & 0 \\ B_{10}^1 & A_1^1 & B_1^1 & \cdots & 0 & 0 \\ 0 & C_1^1 & A_1^1 & \cdots & 0 & 0 \\ \vdots & \vdots & \vdots & \ddots & \vdots & \vdots \\ 0 & 0 & 0 & \cdots & A_1^1 & B_1^1 \\ 0 & 0 & 0 & \cdots & C_1^1 & A_2^1 \end{bmatrix}$$

where A_{00}^1 is the scalar matrix and B_{01}^1 and B_{10}^1 are the row and column vectors of dimensions (1×2) and (2×1), respectively. Similarly, A_1^1, B_1^1, C_1^1, and A_2^1 are the square matrices of order 2. The structures of these sub-block matrices are defined as

$$A_{00}^1 = [-\lambda] \quad B_{01}^1 = [\lambda \quad 0] \quad B_{10}^1 = [\mu_b \quad 0]^T$$

$$A_1^1 = \begin{bmatrix} -(\lambda + \alpha + \mu_b) & \alpha \\ \beta & -(\lambda + \beta) \end{bmatrix}$$

$$B_1^1 = \begin{bmatrix} \lambda & 0 \\ 0 & \lambda \end{bmatrix}$$

$$C_1^1 = \begin{bmatrix} \mu_b & 0 \\ 0 & 0 \end{bmatrix}$$

and

$$A_2^1 = \begin{bmatrix} -(\alpha + \mu_b) & \alpha \\ \beta & -\beta \end{bmatrix}$$

In the next phase, we compute the closed-form expressions of several system performance indicators. These performance indicators are essential in checking any service system's efficiency and working capability. These performance measures may be qualitative or quantitative, supporting system designers and decision-makers in ranking complex service systems. The following are some primary queueing performance measures necessary to explore any queueing-based service system.

- Expected number of customers in the service system at time instant t

$$L_S(t) = \sum\nolimits_{j=0}^{1} \sum\nolimits_{n=j}^{K} n P_{j,n}(t) \tag{3.7}$$

- Expected number of waiting customers in the service system at time instant t

$$L_Q(t) = \sum\nolimits_{j=0}^{1} \sum\nolimits_{n=n}^{K} (n-1) P_{j,n}(t) \tag{3.8}$$

- Probability that the server is idle at time instant t

$$P_I(t) = P_{0,0}(t) \tag{3.9}$$

- Probability that the server is busy in the regular working mode at time instant t

$$P_B(t) = \sum\nolimits_{n=1}^{K} P_{0,n}(t) \tag{3.10}$$

- Probability that the server is broken down at time instant t

$$P_D(t) = \sum\nolimits_{n=1}^{K} P_{1,n}(t) \tag{3.11}$$

- Throughput of the service system at time instant t

$$\tau_p(t) = \sum_{n=1}^{K} \mu_b P_{0,n}(t) \tag{3.12}$$

- Effective arrival rate at time instant t

$$\lambda_{\text{eff}}(t) = \sum_{j=0}^{1} \sum_{n=j}^{K-1} \lambda P_{j,n}(t) \tag{3.13}$$

- Expected waiting time of customer in the service system at time instant t

$$W_s(t) = \frac{L_s(t)}{\lambda_{eff}(t)} \tag{3.14}$$

The numerical simulation is performed in the subsequent section for the economic analysis based on the system performance characteristics above.

3.3 SERVICE SYSTEM WITH WORKING BREAKDOWN OF THE SERVER

Notations

- λ : Mean arrival rate of customers in the service system
- μ_b : Mean service rate during the busy period of the server
- μ_d : Mean service rate when the server is partially broken down
- α : Breakdown rate of the server
- β : Repair rate when the server is broken down
- K : Capacity of the service system

In all the research papers on queue-based service systems with server breakdown, the fundamental assumption has been that a breakdown server terminates the service entirely in the service system. Many authors have studied such a service system with repair as a reliability-based model in the queueing literature. However, a server's breakdown may not completely stop a customer's service in practical situations. For example, the presence of a virus in the system may slow down the computer system's performance. The machine replace problem (MRP) provides another example. When the machine (primary unit) breaks down, it is immediately replaced by another machine (standby unit). The substitute unit works slower until the primary unit is repaired. Therefore, when a failed server provides the service at a lower service rate to the waiting customers in the breakdown state, it is referred to as the server's working breakdown (WB) state.

The basic assumptions opted previously for the finite-capacity service system with server breakdown are also considered here for the study of the service system with WB of the server. Let the service times during the WB state of the server follow an exponential distribution with parameter μ_d . We assume that busy and WB states' inter-arrival and service times are mutually independent.

Let $J(t)$ represent the server's state at time t, and let $N(t)$ represent the total number of customers in the service system at time instant t. Therefore, the possible states of the server are characterized as follows.

$$J(t) = \begin{cases} 0; & \text{The server is in normal working attribute at time instant } t \\ 1; & \text{The server is in WB state at time instant } t \end{cases}$$

Clearly, $\{J(t), N(t)\}$ for $t \geq 0$ represents a continuous-time Markov chain (CTMC) with the state space

$$\Theta_2 \equiv \{(j,n); \ j = 0,1 \ \& \ n = 1,2,\cdots,K\} \cup \{(0,0)\}$$

Using the elementary properties of probability law and balancing the transitions between adjacent states in Figure 3.2 of a finite-capacity service system with WB of the server, the governing differential-difference equations are developed as follows:

- When there is no customer in the service system, and the server is idle

$$\frac{dP_{0,0}(t)}{dt} = -\lambda P_{0,0}(t) + \mu_b P_{0,1}(t) \tag{3.15}$$

- When there are n customers in the service system, and the server is busy in regular working attribute

$$\frac{dP_{0,n}(t)}{dt} = -(\lambda + \mu_b + \alpha) P_{0,n}(t) + \lambda P_{0,n-1}(t)$$
$$+ \mu_b P_{0,n+1}(t) + \beta P_{1,n}(t); 1 \leq n \leq K-1 \tag{3.16}$$

- When the capacity of the system becomes full, and the server is busy with regular working attribute

$$\frac{dP_{0,K}(t)}{dt} = -(\mu_b + \alpha) P_{0,K}(t) + \lambda P_{0,K-1}(t) + \beta P_{1,K}(t) \tag{3.17}$$

- When there is only one customer in the service system, and the server is in the WB state

$$\frac{dP_{1,1}(t)}{dt} = -(\lambda + \beta) P_{1,1}(t) + \alpha P_{0,1}(t) + \mu_d P_{1,2}(t) \tag{3.18}$$

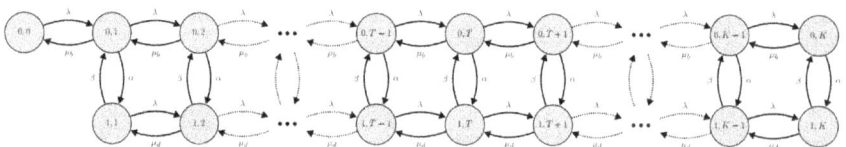

Figure 3.2 State-transition diagram of a finite-capacity service system with working breakdown of the server.

- When there are n customers in the service system, and the server is in the WB state

$$\frac{dP_{1,n}(t)}{dt} = -\left(\lambda + \beta + \mu_d\right)P_{1,n}(t) + \lambda P_{1,n-1}(t) + \alpha P_{0,n}(t)$$
$$+ \mu_d P_{1,n+1}(t); 2 \leq n \leq K - 1 \tag{3.19}$$

- When the capacity of the system becomes full, and the server is in the WB state

$$\frac{dP_{1,K}(t)}{dt} = -\left(\beta + \mu_d\right)P_{1,K}(t) + \lambda P_{1,K-1}(t) + \alpha P_{0,K}(t) \tag{3.20}$$

The generator matrix, denoted by Q_2, is the composition of sub-block matrices obtained by the corresponding transitions between adjacent states of the service system. The structure of the generator matrix is depicted as follows:

$$Q_2 = \begin{bmatrix} A_{00}^2 & B_{01}^2 & 0 & 0 & \cdots & 0 & 0 \\ B_{10}^2 & A_1^2 & B_1^2 & 0 & \cdots & 0 & 0 \\ 0 & C_1^2 & A_2^2 & B_1^2 & \cdots & 0 & 0 \\ 0 & 0 & C_1^2 & A_2^2 & \cdots & 0 & 0 \\ \vdots & \vdots & \vdots & \vdots & \ddots & \vdots & \vdots \\ 0 & 0 & 0 & 0 & \cdots & A_2^2 & B_1^2 \\ 0 & 0 & 0 & 0 & \cdots & C_1^2 & A_3^2 \end{bmatrix}$$

where the block matrix A_{00}^2 is a scalar matrix and B_{01}^2 and B_{10}^2 are the row and column vectors of dimensions (1×2) and (2×1), respectively. Similarly, the sub-block matrices A_1^2, B_1^2, C_1^2, A_2^2, and A_3^2 are the square matrices of order 2. The structures of these sub-matrices are represented as

$$A_{00}^2 = [-\lambda], \quad B_{01}^2 = [\lambda \ \ 0], \quad B_{10}^2 = [\mu_b \ \ 0]^T$$

$$A_1^2 = \begin{bmatrix} -\left(\lambda + \mu_b + \alpha\right) & \alpha \\ \beta & -\left(\lambda + \beta\right) \end{bmatrix}$$

$$A_2^2 = \begin{bmatrix} -\left(\lambda + \mu_b + \alpha\right) & \alpha \\ \beta & -\left(\lambda + \mu_d + \beta\right) \end{bmatrix}$$

$$B_1^2 = \begin{bmatrix} \lambda & 0 \\ 0 & \lambda \end{bmatrix}$$

$$C_1^2 = \begin{bmatrix} \mu_b & 0 \\ 0 & \mu_d \end{bmatrix}$$

and

$$A_3^2 = \begin{bmatrix} -(\mu_b + \alpha) & \alpha \\ \beta & -(\mu_d + \beta) \end{bmatrix}$$

The closed-form expressions for the expected number of customers in the service system $L_S(t)$, the expected number of waiting customers in the service system $L_Q(t)$, the probability that the server is in regular busy mode $P_B(t)$, and the probability that the server is in the WB state $P_{WB}(t)$ are demonstrated in the following manner:

- Expected number of customers in the service system at time instant t

$$L_S(t) = \sum_{j=0}^{1} \sum_{n=j}^{K} n P_{j,n}(t) \tag{3.21}$$

- Expected number of waiting customers in the service system at time instant t

$$L_Q(t) = \sum_{j=0}^{1} \sum_{n=1}^{K} (n-1) P_{j,n}(t) \tag{3.22}$$

- Probability that the server is idle at time instant t

$$P_I(t) = P_{0,0}(t) \tag{3.23}$$

- Probability that the server is busy in the regular working mode at time instant t

$$P_B(t) = \sum_{n=1}^{K} P_{0,n}(t) \tag{3.24}$$

- Probability that the server is in WB state at time instant t

$$P_{WB}(t) = \sum_{n=1}^{K} P_{1,n}(t) \tag{3.25}$$

- Throughput of the service system at time instant t

$$\tau_p(t) = \sum_{n=1}^{K} \mu_b P_{0,n}(t) + \sum_{n=1}^{K} \mu_d P_{1,n}(t) \tag{3.26}$$

- Effective arrival rate at time instant t

$$\lambda_{eff}(t) = \sum_{j=0}^{1} \sum_{n=j}^{K-1} \lambda P_{j,n}(t) \tag{3.27}$$

- Expected waiting time of customer in the service system at time instant t

$$W_s(t) = \frac{L_S(t)}{\lambda_{eff}(t)} \tag{3.28}$$

3.4 SERVICE SYSTEM WITH CUSTOMER IMPATIENCE, WB, SERVICE PRESSURE CONDITION, AND THRESHOLD RECOVERY POLICY

Notations

- λ: Mean arrival rate of customers in the service system
- μ_b^*: Overall service rate during the busy period of the server
- μ_d^*: Overall service rate when the server is partially broken down
- ξ: Balking probability of customers in the service system
- η: Rate of abandonment of customers in the service system
- α: Breakdown rate of the server
- β: Repair rate when the server is broken down
- T : Threshold value for opting recovery policy
- K : Capacity of the service system

This section chooses the queueing nomenclatures and assumptions as in previously defined models. Additionally, some more queueing terminologies, namely customer impatience, service pressure coefficient, and threshold-based recovery policy, have been used to enhance the quality performance of service systems. In general, for the servers who have direct contact with customers, the intensity of the pressure varies with the number of waiting customers in the queue, especially when many customers are waiting for service by limited service providers. However, very few studies have proposed that some employees may perform better under moderate pressure conditions. Hence, we encounter a case in which the unreliable server is under work pressure in his busy period. It is assumed that the service time during the working breakdown state of the server follows an exponential distribution with rate μ_d. During the regular busy period, it is considered that the server provides the service with a constant service rate μ_b as long as the number of customers in the service system, denoted by n, is at most T. When $T \le n \le K$, the service pressure coefficient is considered to make the service rate higher. Therefore, the overall service rate in the typical working attribute (μ_b^*) is defined as

$$\mu_b^* = \begin{cases} \mu_b; & 1 \le n \le T \\ \left[\dfrac{n(T+1)}{T(n+1)} \right]^{\psi} \mu_b; & T+1 \le n \le K \end{cases}$$

In addition, we employ the threshold-based recovery policy along with the WB state of the server. In a threshold-based recovery policy, the server breaks down only when at least one customer is in the service system. At the same time, the recovery can be possible only when $T(1 \le n \le K)$ or more customers are in the system.

Next, by combining all the aforementioned queueing terminologies, we develop a realistic queueing situation that is modeled as follows. Let $N(t)$ be the number of customers in the service system at time instant t, and $J(t)$ be the server's state at time instant t. Then,

$$J(t) = \begin{cases} 0; \text{ the server is in a regular busy period at time instant } t \\ 1; \text{ the server is in WB state at time instant } t \end{cases}$$

Then, $\{(J(t), N(t)); t \geq 0\}$ becomes a continuous-time Markov chain (CTMC) with state space

$$\Theta_3 \equiv \{(j,n); j = 0,1 \text{ and } n = 1,2,\cdots,K\} \cup \{(0,0)\}$$

Figure 3.3 represents the state transition diagram of a queue-based service system with WB of the server and threshold-based recovery policy in a service pressure environment. For mathematical modeling, the governing set of Chapman–Kolmogorov differential-difference equations is given as follows:

- When there is no customer in the service system, and the server is idle

$$\frac{dP_{0,0}(t)}{dt} = -\lambda P_{0,0}(t) + \mu_b^* P_{0,1}(t) \tag{3.29}$$

- When there is only one customer in the service system, and the server is providing the service in regular working attribute

$$\frac{dP_{0,1}(t)}{dt} = -\left(\xi\lambda + \mu_b^* + \alpha\right)P_{0,1}(t) + \lambda P_{0,0}(t) + \left(\eta + \mu_b^*\right)P_{0,2}(t) \tag{3.30}$$

- When there are $n(1 < n < T)$ customers in the service system, and the server is providing the service in regular working attribute

$$\frac{dP_{0,n}(t)}{dt} = -\left(\xi\lambda + (n-1)\eta + \mu_b^* + \alpha\right)P_{0,n}(t)$$

$$+ \xi\lambda P_{0,n-1}(t) + \left(n\eta + \mu_b^*\right)P_{0,n+1}(t); \qquad 2 \leq n \leq T-1 \tag{3.31}$$

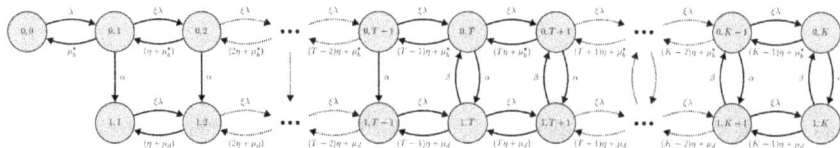

Figure 3.3 State transition diagram of a service system with customer impatience, WB, service pressure condition, and threshold recovery policy.

- When there are $n(\geq T)$ customers in the service system, and the server is providing the service in the regular working attribute

$$\frac{dP_{0,n}(t)}{dt} = -\left(\xi\lambda + (n-1)\eta + \mu_b^* + \alpha\right)P_{0,n}(t) + \xi\lambda P_{0,n-1}(t)$$
$$+ \left(n\eta + \mu_b^*\right)P_{0,n+1}(t) + \beta P_{1,n}(t); \; T \leq n \leq K-1 \qquad (3.32)$$

- When the capacity of the system becomes full, and the server is providing the service in regular working attribute

$$\frac{dP_{0,K}(t)}{dt} = -\left((K-1)\eta + \mu_b^* + \alpha\right)P_{0,K}(t) + \xi\lambda P_{0,K-1}(t) + \beta P_{1,K}(t) \quad (3.33)$$

- When there is only one customer in the service system, and the server is providing the service in the WB environment

$$\frac{dP_{1,1}(t)}{dt} = -\xi\lambda P_{1,1}(t) + \alpha P_{0,1}(t) + \left(\eta + \mu_d\right)P_{1,2}(t) \qquad (3.34)$$

- When there are $n(1 < n < T)$ customers in the service system, and the server is providing the service in the WB environment

$$\frac{dP_{1,n}(t)}{dt} = -\left(\xi\lambda + (n-1)\eta + \mu_d\right)P_{1,n}(t) + \xi\lambda P_{1,n-1}(t)$$
$$+ \alpha P_{0,n}(t) + \left(n\eta + \mu_d\right)P_{1,n+1}(t); \qquad 2 \leq n \leq T-1 \quad (3.35)$$

- When there are $n(\geq T)$ customers in the service system, and the server is providing the service in the WB environment

$$\frac{dP_{1,n}(t)}{dt} = -\left(\xi\lambda + \beta + (n-1)\eta + \mu_d\right)P_{1,n}(t) + \xi\lambda P_{1,n-1}(t)$$
$$+ \alpha P_{0,n}(t) + \left(n\eta + \mu_d\right)P_{1,n+1}(t); \; T \leq n \leq K-1 \qquad (3.36)$$

- When the capacity of the system becomes full, and the server is providing the service in the WB environment

$$\frac{dP_{1,K}(t)}{dt} = -\left(\beta + (K-1)\eta + \mu_d\right)P_{1,K}(t) + \xi\lambda P_{1,K-1}(t) + \alpha P_{0,K}(t) \quad (3.37)$$

Now, using the lexicographic sequence of the states of the service system, the structure of the tri-diagonal transition rate matrix Q_3 is represented as

$$
Q_3 = \begin{bmatrix}
A_{00}^3 & B_{01}^3 & 0 & \cdots & 0 & 0 & 0 & \cdots & 0 & 0 & 0 \\
B_{10}^3 & A_1^3 & B_1^3 & \cdots & 0 & 0 & 0 & \cdots & 0 & 0 & 0 \\
0 & C_1^3 & A_2^3 & \cdots & 0 & 0 & 0 & \cdots & 0 & 0 & 0 \\
\vdots & \vdots & \vdots & \ddots & \vdots & \vdots & \vdots & \ddots & \vdots & \vdots & \vdots \\
0 & 0 & 0 & \cdots & A_{T-1}^3 & B_1^3 & 0 & \cdots & 0 & 0 & 0 \\
0 & 0 & 0 & \cdots & C_{T-1}^3 & A_T^3 & B_1^3 & \cdots & 0 & 0 & 0 \\
0 & 0 & 0 & \cdots & 0 & C_T^3 & A_{T+1}^3 & \cdots & 0 & 0 & 0 \\
\vdots & \vdots & \vdots & \ddots & \vdots & \vdots & \vdots & \ddots & \vdots & \vdots & \vdots \\
0 & 0 & 0 & \cdots & 0 & 0 & 0 & \cdots & A_{K-2}^3 & B_1^3 & 0 \\
0 & 0 & 0 & \cdots & 0 & 0 & 0 & \cdots & C_{K-2}^3 & A_{K-1}^3 & B_1^3 \\
0 & 0 & 0 & \cdots & 0 & 0 & 0 & \cdots & 0 & C_{K-1}^3 & A_K^3
\end{bmatrix}
$$

where the row and column vectors and sub-block matrices are depicted as follows.

$$A_{00}^3 = [-\lambda], \quad B_{01}^3 = [\lambda \ \ 0], \quad B_{10}^3 = [\mu_b^* \ \ 0]^T$$

$$A_1^3 = \begin{bmatrix} -(\xi\lambda + \mu_b^* + \alpha) & \alpha \\ 0 & -\xi\lambda \end{bmatrix}$$

$$A_n^3 = \begin{bmatrix} -(\xi\lambda + (n-1)\eta + \mu_b^* + \alpha) & \alpha \\ 0 & -(\xi\lambda + (n-1)\eta + \mu_d) \end{bmatrix}; 2 \le n \le T-1$$

$$A_n^3 = \begin{bmatrix} -(\xi\lambda + (n-1)\eta + \mu_b^* + \alpha) & \alpha \\ \beta & -(\xi\lambda + (n-1)\eta + \mu_d + \beta) \end{bmatrix}; T \le n \le K-1$$

$$A_K^3 = \begin{bmatrix} -((K-1)\eta + \mu_b^* + \alpha) & \alpha \\ \beta & -((K-1)\eta + \mu_d + \beta) \end{bmatrix}$$

$$B_1^3 = \begin{bmatrix} \xi\lambda & 0 \\ 0 & \xi\lambda \end{bmatrix}$$

and

$$C_n^3 = \begin{bmatrix} n\eta + \mu_b^* & 0 \\ 0 & n\eta + \mu_d \end{bmatrix}; 1 \le n \le K-1$$

Now, to analyze the performance quality and characteristics of the service system, we delineate some system performance measures, namely, the expected number of customers in the system, the system's throughput, and the probability that the server is idle/busy/in the WB state, etc. The closed-form expressions in terms of the transient-state probabilities of these system performance indicators are as follows:

- Expected number of customers in the service system at time instant t

$$L_S(t) = \sum_{j=0}^{1} \sum_{n=j}^{K} n P_{j,n}(t) \tag{3.38}$$

- Expected number of waiting customers in the service system at time instant t

$$L_Q(t) = \sum_{j=0}^{1} \sum_{n=1}^{K} (n-1) P_{j,n}(t) \tag{3.39}$$

- Probability that the server is idle at time instant t

$$P_I(t) = P_{0,0}(t) \tag{3.40}$$

- Probability that the server is busy in the regular working mode at time instant t

$$P_B(t) = \sum_{n=1}^{K} P_{0,n}(t) \tag{3.41}$$

- Probability that the server is in WB state at time instant t

$$P_{WB}(t) = \sum_{n=1}^{K} P_{1,n}(t) \tag{3.42}$$

- Throughput of the service system at time instant t

$$\tau_p(t) = \sum_{n=1}^{K} \mu_b^* P_{0,n}(t) + \sum_{n=1}^{K} \mu_d P_{1,n}(t) \tag{3.43}$$

- Average balking rate in the service system at time instant t

$$ABR(t) = \sum_{j=0}^{1} \sum_{n=1}^{K-1} (1-\xi) \lambda P_{j,n}(t) \tag{3.44}$$

- Average reneging rate in the service system at time instant t

$$ARR(t) = \sum_{j=0}^{1} \sum_{n=1}^{K} (n-1) \eta P_{j,n}(t) \tag{3.45}$$

- Effective arrival rate at time instant t

$$\lambda_{eff}(t) = \lambda P_{0,0}(t) + \sum_{j=0}^{1} \sum_{n=1}^{K-1} (\xi\lambda) P_{j,n}(t) \tag{3.46}$$

- Expected waiting time of customer in the service system at time instant t

$$W_S(t) = \frac{L_S(t)}{\lambda_{eff}(t)} \tag{3.47}$$

3.5 SPECIAL CASES

The following are some exceptional cases of the developed Model 3, already available in the existing literature.

Case 1: On relaxing the service pressure condition and customer impatience, i.e., $\mu_b^* = \mu_b$, $\mu_d^* = \mu_d$, $\xi = 1$, and $\eta = 0$, the results obtained in the transient state of the developed model match with the article [42].

Case 2: In addition to special case 1, if $\mu_d = 0$ also, the studied queueing model is similar to [34].

Case 3: On relaxing the customer impatience, service pressure condition, and threshold-based recovery policy, the given model is equivalent to [18].

Case 4: If $T = K \rightarrow \infty$, $\xi = 1$, $\eta = 0$, and $\mu_d = 0$, the model deduces to $M/M/1$ queueing system with the unreliable server (cf. [43]).

Case 5: If $T = K \rightarrow \infty$, $\alpha = \beta = \eta = 0$, and $\mu_d = 0$, the model becomes $M/M/1$ queueing system with balking (cf. [44]).

Case 6: For $T = K \rightarrow \infty$, $\alpha = \beta = 0$, $\xi = 1$, and $\mu_d = 0$, the investigated model is similar to $M/M/1$ queueing system with reneging (cf. [45]).

Case 7: The governing model can further be extended for $M/M/c$ queueing model with balking and reneging by setting $T = K \rightarrow \infty$, $\alpha = \beta = 0$, and $\mu_d = 0$ (cf. [46]).

3.6 COST ANALYSIS

This section develops the governing service system's expected total cost function to formulate the cost optimization problem. To calculate optimal system design parameters μ_b^{**} and μ_d^{**}, it helps system analysts and engineers in decision-making.

3.6.1 Steady-state analysis

In this subsection, the steady-state analysis of the developed queue-based service system at equilibrium is performed to examine the optimal operating policy. In equilibrium, i.e., $t \rightarrow \infty$, the steady-state probabilities of the governing system are defined as follows.

$$P_{0,0} = \lim_{t \to \infty} Pr\big[J(t) = 0, N(t) = 0\big]$$

and

$$P_{j,n} = \lim_{t \to \infty} Pr\big[J(t) = j, N(t) = n\big]; \quad j = 0,1 \quad \& \quad n = 1,2,\cdots,K$$

Now, using the matrix method based on the matrix equation $Q_3 Y = 0$, the steady-state probability distribution can be easily demonstrated. Furthermore, for optimal analysis, the expected cost function is also depicted using intrinsic system performance indicators in the following subsection that incur some default costs.

3.6.2 Cost function

For the economic analysis, the system design parameters, namely, μ_b (service rate during the busy period) and μ_d (service rate during the WB state), are taken into consideration. In this chapter, the main objective of our intuition is to illustrate the optimal service rates, say μ_b^{**} and μ_d^{**} in busy and WB states, respectively, for minimizing the expected total cost of the service system. Following are some associated cost elements of different system performance indicators and system design parameters that are considered and defined as follows:

$C_h \equiv$ Holding cost for each customer present in the service system

$C_b \equiv$ The cost associated with the regular busy state of the server

$C_{wb} \equiv$ The cost associated with the WB state of the server

$C_i \equiv$ Fixed cost for the idle state of the server

$C_m \equiv$ The associated cost for providing the service with rate μ_b

$C_d \equiv$ The associated cost for providing the service with rate μ_d

Using the queueing-theoretic approach and the aforementioned cost elements, the cost function is formulated as

$$TC(\mu_b, \mu_d) = C_h L_S + C_b P_B + C_{wb} P_{WB} + C_i P_I + C_m \mu_b + C_d \mu_d \qquad (3.48)$$

The values of the system parameters λ and α in a service system are not usually to be controlled due to the outside factors of the system. For some purposes, the decision variables μ_b and μ_d can be viewed as controllable variables. Therefore, the cost minimization problem of the described model involved in the service system with WB, pressure coefficient, and threshold-based recovery policy can be mathematically represented as a constrained problem as follows:

$$TC(\mu_b^{**}, \mu_d^{**}) = \min_{(\mu_b, \mu_d)} TC(\mu_b, \mu_d) \qquad (3.49)$$

$$s.t. \quad \mu_d < \mu_b$$

In the upcoming section, we reduce the expected cost (TC) by adjusting μ_b and μ_d under the constraint $\mu_d < \mu_b$, which minimizes the total cost function (48). In the cost function (48), two decision variables μ_b and μ_d are included, where both variables are continuous quantities. The analytic development of the optimal solution $\left(\mu_b^{**}, \mu_d^{**}\right)$ is arduous due to the optimization problem's highly non-linear and complex nature. As a result, it is not easy to show the convexity of the developed cost function. Therefore, to determine the optimal pair $\left(\mu_b^{**}, \mu_d^{**}\right)$, we dealt with the continuous optimization variables $\left(\mu_b, \mu_d\right)$ and employed the Quasi-Newton algorithm in the next section.

3.6.3 Quasi-Newton method

The quasi-Newton method is an efficient optimization technique for finding the optimum (minimum) of non-linear functions. This method is generally employed to decide the search direction with an iterative process. These iterations are performed until a tolerance level is achieved by trying different step lengths in this direction for a better optimal solution. The classical optimization technique, the Quasi-Newton method, uses a second-order derivative as a Hessian matrix and provides appreciably better solutions when the solution is closer to the initial point than the gradient methods. In addition, the solution process of the Quasi-Newton method is more transparent than other optimization algorithms. Hence, we apply the Quasi-Newton optimization technique to obtain optimal global pair $\left(\mu_b^{**}, \mu_d^{**}\right)$ when all the other systems parameters are initially fixed. An iterative procedure based on the Quasi-Newton method is executed as follows.

Ω Quasi-Newton method: Pseudo-code

Input: Input parameters, the initial value $\Omega_{(0)} = \left[\mu_b^{(0)}, \mu_d^{(0)}\right]^T$, and tolerance ε.

Output: Approximate the solution $[\mu_b^{**}, \mu_d^{**}]^T$ and ascertain the subsequent value of the cost function $TC\left(\mu_b^{**}, \mu_d^{**}\right)$.

Step 1: Introduce the initial trial solution $\Omega_{(0)}$ and determine $TC\left(\Omega_{(0)}\right)$.

Step 2: while $\left|\dfrac{\partial TC}{\partial \mu_b}\right| > \varepsilon$ or $\left|\dfrac{\partial TC}{\partial \mu_d}\right| > \varepsilon$, **do** steps 3 and 4.

Step 3: Enumerate the gradient of the expected cost function

$$\vec{\nabla}TC(\Omega) = \left[\frac{\partial TC}{\partial \mu_b}, \frac{\partial TC}{\partial \mu_d}\right]^T.$$ Also, figure out the Hessian matrix

$$H(\Omega) = \begin{bmatrix} \dfrac{\partial^2 TC}{\partial \mu_b^2} & \dfrac{\partial^2 TC}{\partial \mu_b \, \partial \mu_d} \\ \dfrac{\partial^2 TC}{\partial \mu_d \, \partial \mu_b} & \dfrac{\partial^2 TC}{\partial \mu_d^2} \end{bmatrix} \text{ at point } \vec{\Omega}_i.$$

Step 4: Update the solution $\Omega_{(i+1)} = \Omega_{(i)} - \left[H\left(\Omega_{(i)}\right) \right]^{-1} \vec{\nabla} TC\left(\Omega_{(i)}\right)$. **end**
Step 5: Output

Readers can review the papers (*cf.* [47], [48], [49], [50], [51], [52], [53], [54], [55], [56], [57], [58], [59]), and references cited therein for more detail regarding optimal strategies, analytic solution techniques, and global search optimization techniques.

3.7 NUMERICAL RESULTS AND DISCUSSION

The main objective of this study in this chapter is to understand the qualitative and perceptible performance of the established queue-based service system using the server's working breakdown and threshold-based recovery policy. This section illustrates some numerical experiments and provides a straightforward comparative analysis based on three studied models.

Model 1: Single-server service system with server breakdown
Model 2: Single-server service system with WB of the server
Model 3: Service system with WB, service pressure condition, and threshold recovery policy of the server

For numerical investigation, first, we fix the default values of the system parameters as $K = 15$, $T = 7$, $\lambda = 1.75$, $\xi = 0.9$, $\eta = 0.005$, $\psi = 1.0$, $\alpha = 0.1$, and $\beta = 3.0$ along with the unit cost elements as $C_b = 15$, $C_{wb} = 135$, $C_b = 250$, $C_i = 170$, $C_m = 5$, and $C_d = 10$, respectively. With the aid of MATLAB (2018b) software, all the numerical research findings are demonstrated and depicted with the help of different graphs (*cf.* Figures 3.4–3.6) and tables (*cf.* Tables 3.1 and 3.2).

For the cost analysis, first, we depict the optimal operating strategies (i.e., optimal values of system design parameters μ_b and μ_d with respect to optimum expected cost) with the help of convex plots in Figure 3.4(i) and (ii). Next, for the optimal analysis, we provide a surface plot for the expected total cost function (48) in Figure 3.5 for combined values of service system design parameters μ_b and μ_d besides the aforementioned system parameter value and unit cost elements. Figure 3.5 convinces us of the convex

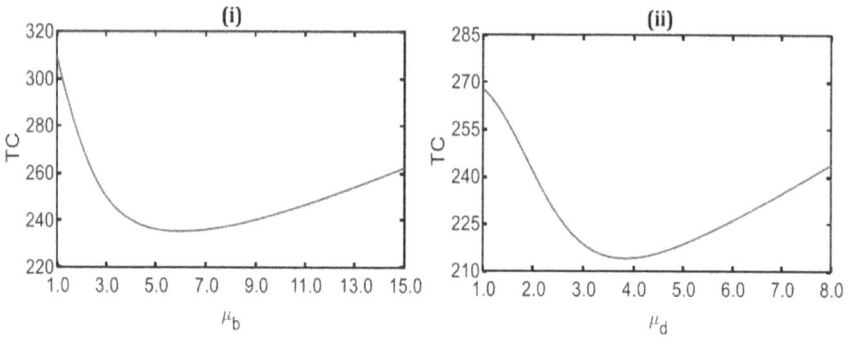

Figure 3.4 Optimal values of system design parameters μ_b and μ_d along with optimal cost TC.

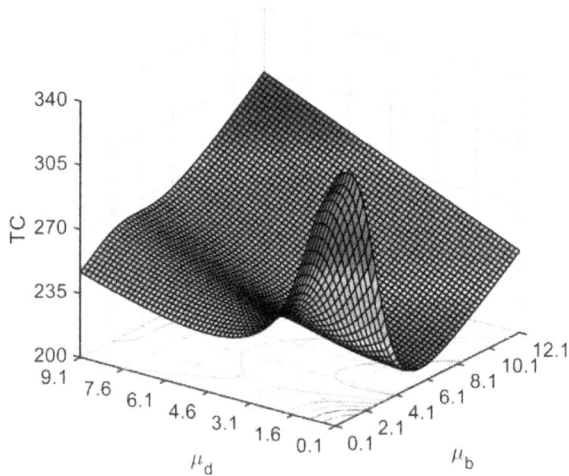

Figure 3.5 Surface plot for the expected total cost (TC) with respect to system design parameters μ_b and μ_d.

shape of the surface plot, which validates that the formulated cost function is unimodal (convex) in nature. Hence, a unique relative minimum of the expected cost function exists for a specific combination of design parameters μ_b and μ_d.

Furthermore, the comparative analysis among all three developed models based on the expected cost of the service system is provided with the help of Figure 3.6. In Figure 3.6(i), it is noted that as the service rate during the busy period of the server μ_b Increases, the expected cost for all the developed models decreases for some instant and then increases continuously. This trend convinces us that the desired cost function is convex in nature, as per

(i)

(ii)

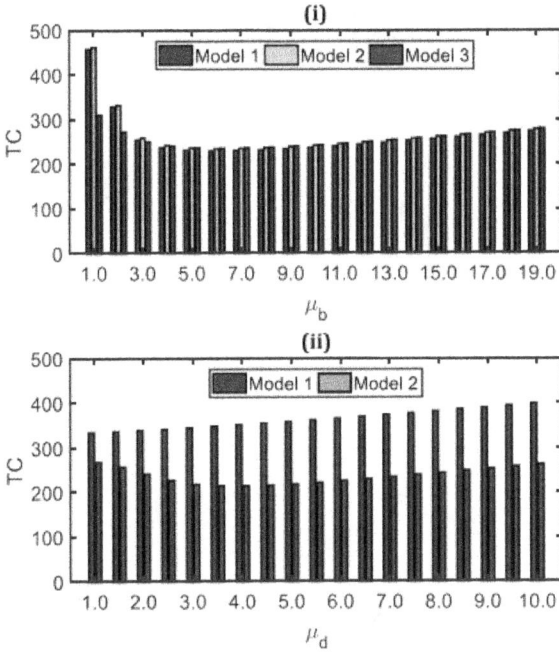

Figure 3.6 Comparative analysis of the governing models 1, 2, and 3 based on the expected cost of the system *(TC)* with respect to system design parameters μ_b and μ_d.

our intuition. Similarly, Figure 3.6(ii) prompts that, as the service rate during the working breakdown state μ_d increases, the expected cost increases continuously for model 2 and becomes convex in the case of model 3. Therefore, as a concluding remark, we can say that it is not always advisable to provide a higher service rate to get better service from the service system.

Subsequently, an economic analysis is performed to achieve the combined optimal values $\left(\mu_b^{**}, \mu_d^{**}\right)$ by employing the semi-classical optimization technique, the Quasi-Newton method. For this purpose, we select the values of the default system parameters as mentioned earlier and the initial trial solution $\left(\mu_b^{(0)}, \mu_d^{(0)}\right) = (4.5, 3.0)$ with the initial cost value $TC\left(\mu_b^{(0)}, \mu_d^{(0)}\right) = 227.187356$. After seven iterations, Table 3.1 reveals that the minimum expected cost of the service system converges with the solution $\left(\mu_b^{**}, \mu_d^{**}\right) = (4.838052, 3.088053)$ with the value 228.239565. Next, we utilize the result of Table 3.1 and perform the optimal analysis for several other combinations of system parameters λ, α, and β by the Quasi-Newton method in Table 3.2. Table 3.2 prompts that better service rates in both busy and WB states are required for more customers in the service system. It allows the service rate to be a little less in the WB state, which signifies the nonworking attitude of the server in WB mode. Furthermore,

Table 3.1 Quasi-Newton Algorithm in Searching for the Optimal Solution $\left(\mu_b^{**}, \mu_d^{**}\right)$, Along With Optimal Expected Cost $TC\left(\mu_b^{**}, \mu_d^{**}\right)$

No. of iterations	0	1	2	3	4	5	6	7
$TC\left(\mu_b, \mu_d\right)$	227.187356	227.266201	228.364279	228.242831	228.240564	228.239575	228.239565	**228.239565**
μ_d	3.0	3.070676	3.083358	3.086364	3.087776	3.088052	3.088053	**3.088053**
μ_b	4.5	4.603968	4.860782	4.837079	4.837995	4.838054	4.838053	**4.838052**

Table 3.2 Quasi-Newton Algorithm in Searching for the Solution $\left(\mu_b^{**}, \mu_d^{**}\right)$ Along With Optimal Expected Cost $TC\left(\mu_b^{**}, \mu_d^{**}\right)$

(λ, α, β)	(1.75, 0.1, 3.0)	(1.5, 0.1, 3.0)	(1.0, 0.1, 3.0)	(1.75, 0.3, 3.0)	(1.75, 0.5, 3.0)	(1.75, 0.1, 4.0)	(1.75, 0.1, 5.0)	(1.75, 0.1, 4.5)
$\left(\mu_b^{(0)}, \mu_d^{(0)}\right)$	(4.5, 3.0)	(4.5, 3.0)	(4.5, 3.0)	(4.5, 3.0)	(4.5, 3.0)	(4.5, 3.0)	(4.5, 3.0)	(4.5, 3.0)
μ_d^{**}	3.088053	2.743741	2.004112	2.726051	2.609750	3.091932	3.093646	3.092962
μ_b^{**}	4.838052	4.493742	3.754126	4.476053	4.359751	4.841934	4.843648	4.842961
$TC\left(\mu_b^{**}, \mu_d^{**}\right)$	228.239565	220.585731	204.504218	221.847226	219.911044	228.701801	229.020868	228.874532

we conclude that the semi-classical optimizer, the Quasi-Newton method is quite helpful and easy to implement in finding the optimal combination $\left(\mu_b^{**}, \mu_d^{**} \right)$ with the minimal expected cost TC^* of the service system.

3.8 CONCLUSION

This chapter investigated a finite-capacity service system with customer impatience, service pressure conditions, and working breakdown under the regime of threshold-based recovery policy of the server. We have developed three different queueing models by combining the aforementioned queueing nomenclatures and provided a comparative analysis for a better understanding from the research point of view. Each developed model has provided the governing set of Chapman–Kolmogorov differential-difference equations. The stationary queue-size distribution has been obtained by employing the matrix approach. Afterward, using the queue-size distribution, several system performance indicators, such as the expected number of customers in the service system, the probability that the server is busy, and the throughput of the service system, are computed. In addition, the cost optimization (minimization) problem has been formulated to determine the optimal operating policy with the minimal expected cost of the service system. For the economic analysis, the semi-classical optimization approach, the Quasi-Newton method, has been implemented, and research findings have been illustrated with the help of various graphs and tables. Based on the numerical experiments, it has been concluded that the Quasi-Newton method is faster in searching for the combined optimal values of the resource decision variables (μ_b^{**} and μ_d^{**}) along with the cost minimization policy.

Finally, as a concluded remark, it can be outlined that our investigations and suggested numerical iterative-search approach help system designers, decision-makers, and queueing theorists to design several multi-server service systems in different real-time scenarios and their complex maintenance mechanisms. Furthermore, the present study and developed models could be extended for the bulk arrival/service, general arrival/service distribution, and equilibrium balking strategies.

Conflict of interest: The authors declare that there is no conflict of interest regarding the publication of this chapter.

REFERENCES

[1] D. P. Gaver Jr., "A waiting line with interrupted service, including priorities," *Journal of the Royal Statistical Society: Series B (Methodological)*, Vol. 24, No. 1, pp. 73–90, 1962.

[2] A. W. Shogan, "A single server queue with arrival rate dependent on server breakdowns," *Naval Research Logistics Quarterly*, Vol. 26, No. 3, pp. 487–497, 1979.

[3] D. Jayaraman, R. Nadarajan, and M. R. Sitrarasu, "A general bulk service queue with arrival rate dependent on server breakdowns," *Applied Mathematical Modelling*, Vol. 18, No. 3, pp. 156–160, 1994.

[4] J. C. Ke, "Bi-level control for batch arrival queues with an early startup and un-reliable server," *Applied Mathematical Modelling*, Vol. 28, No. 5, pp. 469–485, 2004.

[5] M. Jain and P. K. Agrawal, "Optimal policy for bulk queue with multiple types of server breakdown," *International Journal of Operational Research*, Vol. 4, No. 1, pp. 35–54, 2009.

[6] K. H. Wang and D. Y. Yang, "Controlling arrivals for a queueing system with an unreliable server: Newton-quasi method," *Applied Mathematics and Computation*, Vol. 213, No. 1, pp. 92–101, 2009.

[7] C. J. Singh, M. Jain, and B. Kumar, "Analysis of queue with two phases of service and m-phases of repair for server breakdown under N-policy, " *International Journal of Services and Operations Management*, Vol. 16, No. 3, pp. 373–406, 2013.

[8] C. J. Chang, F. M. Chang, and J. C. Ke, "Economic application in a Bernoulli F-policy queueing system with server breakdown," *International Journal of Production Research*, Vol. 52, No. 3, pp. 743–756, 2014.

[9] M. Jain, C. Shekhar, and V. Rani, " N-policy for a multi-component machining system with imperfect coverage, reboot and unreliable server," *Production & Manufacturing Research*, Vol. 2, No. 1, pp. 457–476, 2014.

[10] M. Jain, C. Shekhar, and S. Shukla, "Machine repair problem with an unreliable server and controlled arrival of failed machines," *OPSEARCH*, Vol. 51, No. 3, pp. 416–433, 2014.

[11] M. Zhang and Q. Liu, "An $M/G/1$ G-queue with server breakdown, working vacations and vacation interruption," *OPSEARCH*, Vol. 52, No. 2, pp. 256–270, 2015.

[12] C. C. Kuo and J. C. Ke, "Comparative analysis of standby systems with unreliable server and switching failure," *Reliability Engineering & System Safety*, Vol. 145, pp. 74–82, 2016.

[13] G. Choudhury and M. Deka, "A batch arrival unreliable server delaying repair queue with two phases of service and Bernoulli vacation under multiple vacation policy," *Quality Technology & Quantitative Management*, Vol. 15, No. 2, pp. 157–186, 2018.

[14] F. M. Chang, T. H. Liu, and J. C. Ke, "On an unreliable-server retrial queue with customer feedback and impatience," *Applied Mathematical Modelling*, Vol. 55, pp. 171–182, 2018.

[15] A. Nazarov, J. Sztrik, A. Kvach, and T. Bérczes, "Asymptotic analysis of finite-source $M/M/1$ retrial queueing system with collisions and server subject to breakdowns and repairs," *Annals of Operations Research*, Vol. 277, No. 2, pp. 213–229, 2019.

[16] J. C. Ke, T. H. Liu, S. Su, and Z. G. Zhang, "On retrial queue with customer balking and feedback subject to server breakdowns," *Communications in Statistics-Theory and Methods*, Vol. 51, No. 17, pp. 6049–6063, 2022.

[17] A. Nazarov, J. Sztrik, A. Kvach, and A. Tóth, "Asymptotic analysis of finite source $M/GI/1$ retrial queueing systems with collisions and server subject to breakdowns and repairs," *Methodology and Computing in Applied Probability*, Vol. 24, pp. 1503–1518, 2022.

[18] K. Kalidass and R. Kasturi, "A queue with working breakdowns," *Computers & Industrial Engineering*, Vol. 63, No. 4, pp. 779–783, 2012.

[19] Z. Liu and Y. Song, "The $M^X/M/1$ queue with working breakdown," *RAIRO-Operations Research*, Vol. 48, No. 3, pp. 399–413, 2014.

[20] D. Y. Yang and Y. Y. Wu, "Transient behavior analysis of a finite capacity queue with working breakdowns and server vacations," *Proceedings of the International Multi-Conference of Engineers and Computer Scientists*, Vol. 2, 2014.

[21] G. N. Purohit, M. Jain, and S. Rani, "$M/M/1$ retrial queue with constant retrial policy, unreliable server, threshold based recovery and state dependent arrival rates," *Applied Mathematical Sciences*, Vol. 6, No. 37, pp. 1837–1846, 2012.

[22] R. Sharma and G. Kumar, "Unreliable server $M/M/1$ queue with priority queueing system," *International Journal of Engineering and Technical Research*, Vol. 2014, pp. 368–371, 2014.

[23] C. D. Liou, "Markovian queue optimisation analysis with an unreliable server subject to working breakdowns and impatient customers," *International Journal of Systems Science*, Vol. 46, No. 12, pp. 2165–2182, 2015.

[24] D. Y. Yang and Y. H. Chen, "Computation and optimization of a working breakdown queue with second optional service," *Journal of Industrial and Production Engineering*, Vol. 35, No. 3, pp. 181–188, 2018.

[25] T. Jiang and B. Xin, "Computational analysis of the queue with working breakdowns and delaying repair under a Bernoulli-schedule-controlled policy," *Communications in Statistics Theory and Methods*, Vol. 48, No. 4, pp. 926–941, 2019.

[26] M. Zhang and S. Gao, "The disasters queue with working breakdowns and impatient customers," *RAIRO-Operations Research*, Vol. 54, No. 3, pp. 815–825, 2020.

[27] D. Y. Yang, C. H. Chung, and C. H. Wu, "Sojourn times in a Markovian queue with working breakdowns and delayed working vacations," *Computers & Industrial Engineering*, Vol. 156, p. 107239, 2021.

[28] K. A. Alnowibet, A. F. Alrasheedi, and F. S. Alqahtani, "Queuing models for analyzing the steady-state distribution of stochastic inventory systems with random lead time and impatient customers," *Processes*, Vol. 10, No. 4, p. 624, 2022.

[29] E. Danilyuk, A. Plekhanov, S. Moiseeva, and J. Sztrik, "Asymptotic diffusion analysis of retrial queueing system $M/M/1$ with impatient customers, collisions and unreliable servers," *Axioms*, Vol. 11, No. 12, p. 699, 2022.

[30] X. Chai, T. Jiang, L. Li, W. Xu, and L. Liu, "On a many-to-many matched queueing system with flexible matching mechanism and impatient customers," *Journal of Computational and Applied Mathematics*, Vol. 416, p. 114573, 2022.

[31] E. Bolandifar, N. DeHoratius, and T. Olsen, "Modeling abandonment behavior among patients," *European Journal of Operational Research*, Vol. 306, No. 1, pp. 243–254, 2023.

[32] D. V. Efrosinin and O. V. Semenova, "An $M/M/1$ system with an unreliable device and a threshold recovery policy," *Journal of Communications Technology and Electronics*, Vol. 55, No. 12, pp. 1526–1531, 2010.

[33] D. Efrosinin and A. Winkler, "Queueing system with a constant retrial rate, non-reliable server and threshold-based recovery," *European Journal of Operational Research*, Vol. 210, No. 3, pp. 594–605, 2011.

[34] D. Y. Yang, Y. C. Chiang, and C. S. Tsou, "Cost analysis of a finite capacity queue with server breakdowns and threshold-based recovery policy," *Journal of Manufacturing Systems*, Vol. 32, No. 1, pp. 174–179, 2013.

[35] M. Jain, "Priority queue with batch arrival, balking, threshold recovery, unreliable server and optimal service," *RAIRO-Operations Research*, Vol. 51, No. 2, pp. 417–432, 2017.

[36] Y. Barron, "A threshold policy in a Markov-modulated production system with server vacation: The case of continuous and batch supplies," *Advances in Applied Probability*, Vol. 50, No. 4, pp. 1246–1274, 2018.

[37] V. Poongothal, P. Godhandaraman, and V. Saravanan, "An unreliable $M/M/1$ retrial queue with discouragement under threshold recovery policy," *Recent Advancement of Mathematics in Science and Technology*, p. 125, 2021.

[38] C. H. Lan, C. C. Chang, M. P. Kuo, *et al.*, "Service system under service pressure by system dynamics model," *ProbStat Forum*, Vol. 3, pp. 52–64, 2010.

[39] J. C. Ke, Y. L. Hsu, T. H. Liu, and Z. G. Zhang, "Computational analysis of machine repair problem with unreliable multi-repairmen," *Computers & Operations Research*, Vol. 40, No. 3, pp. 848–855, 2013.

[40] T. H. Liu and J. C. Ke, "On the multi-server machine interference with modified Bernoulli vacation," *Journal of Industrial & Management Optimization*, Vol. 10, No. 4, pp. 1191–1208, 2014.

[41] A. Krishnamoorthy, P. K. Pramod, and S. R. Chakravarthy, "Queues with interruptions: A survey," *TOP*, Vol. 22, No. 1, pp. 290–320, 2014.

[42] N. J. Ezeagu, G. O. Orwa, and M. J. Winckler, "Transient analysis of a finite capacity $M/M/1$ queuing system with working breakdowns and recovery policies," *Global Journal of Pure and Applied Mathematics*, Vol. 14, No. 8, pp. 1049–1065, 2018.

[43] M. Yadin and P. Naor, "Queueing systems with a removable service station," *Journal of the Operational Research Society*, Vol. 14, No. 4, pp. 393–405, 1963.

[44] F. A. Haight, "Queueing with balking," *Biometrika*, Vol. 44, No. 3–4, pp. 360–369, 1957.

[45] F. A. Haight, "Queueing with reneging," *Metrika*, Vol. 2, No. 1, pp. 186–197, 1959.

[46] M. Abou-El-Ata and A. Hariri, "The $M/M/c/N$ queue with balking and reneging," *Computers & Operations Research*, Vol. 19, No. 8, pp. 713–716, 1992.

[47] C. Shekhar, S. Varshney, and A. Kumar, "Optimal control of a service system with emergency vacation using Bat algorithm," *Journal of Computational and Applied Mathematics*, Vol. 364, p. 112332, 2020.

[48] C. Shekhar, S. Varshney, and A. Kumar, "Optimal and sensitivity analysis of vacation queueing system with F-policy and vacation interruption," *Arabian Journal for Science and Engineering*, Vol. 45, No. 8, pp. 7091–7107, 2020.

[49] C. Shekhar, S. Varshney, and A. Kumar, "Matrix-geometric solution of multi-server queueing systems with Bernoulli scheduled modified vacation and

retention of reneged customers: A meta-heuristic approach," *Quality Technology & Quantitative Management*, Vol. 18, No. 1, pp. 39–66, 2021.

[50] C. Shekhar, S. Varshney, and A. Kumar, "Standbys provisioning in machine repair problem with unreliable service and vacation interruption," in *The Handbook of Reliability, Maintenance, and System Safety Through Mathematical Modeling*, pp. 101–133, Elsevier, 2021.

[51] C. Shekhar, P. Deora, S. Varshney, K. P. Singh, and D. C. Sharma, "Optimal profit analysis of machine repair problem with repair in phases and organizational delay," *International Journal of Mathematical, Engineering and Management Sciences*, Vol. 6, No. 1, pp. 442–468, 2021.

[52] C. Shekhar, N. Kumar, A. Gupta, A. Kumar, and S. Varshney, "Warm-spare provisioning computing network with switching failure, common cause failure, vacation interruption, and synchronized reneging," *Reliability Engineering & System Safety*, Vol. 199, p. 106910, 2020.

[53] C. Shekhar, S. Varshney, and A. Kumar, "Reliability and vacation: The critical issue," in *Advances in Reliability Analysis and its Applications*, pp. 251–292, Springer, 2020.

[54] C. Shekhar, A. Kumar, and S. Varshney, "Modified Bessel series solution of the single server queueing model with feedback," *International Journal of Computing Science and Mathematics*, Vol. 10, No. 3, pp. 313–326, 2019.

[55] C. Shekhar, A. Gupta, N. Kumar, A. Kumar, and S. Varshney, "Transient solution of multiple vacation queue with discouragement and feedback," *Scientia Iranica*, Vol. 29, No. 5, 2567–2577, 2022.

[56] C. Shekhar, A. Kumar, and S. Varshney, "Load sharing redundant repairable systems with switching and reboot delay," *Reliability Engineering & System Safety*, Vol. 193, p. 106656, 2020.

[57] C. Shekhar, A. Kumar, S. Varshney, and S. I. Ammar, "Fault-tolerant redundant repairable system with different failures and delays," *Engineering Computations*, Vol. 37, No. 3, pp. 1043–1071, 2019.

[58] C. Shekhar, A. Kumar, S. Varshney, and S. I. Ammar, "$M/G/1$ fault-tolerant machining system with imperfection," *Journal of Industrial & Management Optimization*, Vol. 17, No. 1, p. 1, 2021.

[59] C. Shekhar, A. Kumar, and S. Varshney, "Parametric nonlinear programming for fuzzified queuing systems with catastrophe," *International Journal of Process Management and Benchmarking*, Vol. 10, No. 1, pp. 69–98, 2020.

Chapter 4

Multi-objective fuzzy linear programming problem with fuzzy decision variables

A geometrical approach

Admasu Tadesse, M. M. Acharya,
Manoranjan Sahoo, and Srikumar Acharya

4.1 INTRODUCTION

Mathematical modeling of optimization problems in fuzzy environment is called fuzzy programming problem. One of the most frequently applied fuzzy decision optimization technique is fuzzy linear programming (FLP) problem. Although it has been investigated and expanded for more than decades by many researchers and from the various point of views, it is still useful to develop new approaches in order to better fit the real-world problems within the framework of FLP problem. In particularly when the information is vague, relating to human language and behavior, imprecise/ambiguous system data, or when the information could not be described and defined well due to limited knowledge and deficiency in its understanding, the decision-maker model the optimization problem as fuzzy programming problem. Such types of uncertainty cannot be dealt and solved effectively by traditional mathematics based optimization techniques.

In most of the optimization models, the parameters are assumed to be definite and precisely known. But such assumptions are not suitable to deal with several real-life problems where many parameters are imprecise and vague. Bellman and Zadeh (1970) introduced the concept of decision-making in fuzzy environment. After his innovative work, many researchers have introduced different strategies for solving FLP problems. We can classify FLP problems as partially FLP problem and fully FLP problem according to existing literatures. FLP problems are said to be partially FLP problems if not all parts of the problem are fuzzy and if all the parameter and decisions variables of FLP are fuzzy, then the programming problem is called fully FLP problem as it was named by Buckley and Feuring (2000), for the first time. The researchers, namely, Buckley (1988, 1989), Julien (1994), Maleki (2003), and Inuiguchi et al. (1990) assumed that parameters are fuzzy numbers while the decision variables are crisp ones. Since the variables are crisp, we always obtain the crisp and exact solutions, whose

DOI: 10.1201/9781003462422-4

values are real in the FLP problems with fuzzy parameters as a result the fuzzy aspect of the decision is partly lost, so it is reasonable and important to consider the FLP problems with fuzzy decision variables. The importance and the applications of fuzzy decision variables are well acknowledged by Acharya and Biswal (2015); in their work, they have discussed applications of multi-choice FLP problem to a garment manufacture company, where some of the parameters and the decisions variables are fuzzy numbers. Fully FLP problem has many different applications in sciences and engineering and various methods have been proposed for solving it. Due to the types of constraints, fully fuzzy LP problems can be divided into two parts as fully FLP problems with equality constraints and fully FLP with inequality constraints. Kumar et al. (2011) proposed a method for solving fully FLP problem with equality constraints.

Authors, namely, Allahviranloo et al. (2008), Hashemi et al. (2006), and Nasseri et al. (2013) introduced fully FLP Problems with inequality constraints. Mitlif (2019) introduced a new ranking function of triangular fuzzy numbers for solving FLP problems in objective function and find the optimal solution of it by Big-M method, where decision variables and objective functions coefficients are triangular fuzzy number. Das (2017) proposed modified algorithm and has made some comparative study with existing methods. Recently, Khalifa (2019) proposed FLP problem in which all parameters and variables are characterized by L-R fuzzy numbers. Very recently, Deshmukh et al. (2020) introduced FLP having all the parameters and decision variables as symmetric hexagonal fuzzy number. Moreover, strategies for solving multi-objective FLP problem were also discussed, for instance, Sharma and Aggarwal (2018) introduced fully fuzzy multi-objective linear programming problem in which all the coefficients and decision variables are L-R flat fuzzy numbers. In their work, they considered all the constraint to be fuzzy inequalities constraints. In the same year, Hamadameen and Hassan (2018) proposed compromised solution for the fully fuzzy multi-objective LP problem. Jalil et al. (2017) introduced a solution approach for obtaining compromise optimal solution of fully fuzzy multi-objective solid transportation problem, where the fuzzy objective functions and equality constraints are defuzzified by applying the ranking function and the property of equality between fuzzy numbers, respectively. In their work, the crisp multi-objective problem is then converted into single-objective problem by using weighted sum approach.

We propose a method for solving multi-objective FLP problem, where all the parameters and decision variables are triangular fuzzy number. Weighted sum method is used to convert fuzzy multi-objective linear programming problem into single-objective FLP problem. FLP problems cannot be solved directly. So, we convert this FLP into its equivalent crisp linear programming problem. The concept of incenter of the triangle is used to convert FLP problem into its equivalent crisp linear programming problem. The rest of this chapter is organized as follows. In Section 4.2, some basic definitions on incenter of a triangle and relationship between fuzzy number and vertex points of a triangle are presented. Moreover, definition on the triangular fuzzy number and arithmetic operations between two triangular fuzzy numbers are discussed further in this section. In Section 4.3, general mathematical model is formulated. The main results obtained in this chapter will be clarified by an illustrative numerical example in Section 4.5 followed by conclusion in Section 4.6 with supportive references.

4.2 BASIC PRELIMINARIES

Definition 2.1 A triangular fuzzy number $\tilde{A}$ is denoted by $\left(A^{(p)}, A, A^{(o)}\right)$, where $A^{(p)}, A$, and $A^{(o)}$ are real numbers. The membership function $(\mu_{\tilde{A}}(x))$ of $\tilde{A}$ is given as:

$$\mu_{\tilde{A}}(x) = \begin{cases} 0, & x \leq A^{(p)} \\ \dfrac{x - A^{(p)}}{A - A^{(p)}}, & A^{(p)} \leq x \leq A \\ \dfrac{A^{(o)} - x}{A^{(o)} - A}, & A \leq x \leq A^{(o)} \\ 0 & \text{otherwise} \end{cases}.$$

Definition 2.2 Let $\Delta A'BC$ be triangle with vertex $A'\left(A_x, A_y\right), B\left(B_x, B_y\right)$, and $C(C_x, C_y)$ having opposite side lengths a, b, and c, respectively, then the incenter of the triangle is given by

$$(x', y') = \left(\frac{aA_x + bB_x + cC_x}{P}, \frac{aA_y + bB_y + cC_y}{P}\right)$$

where $P = a + b + c$ is perimeter of the triangle.

Definition 2.3 Let $O(0,0)$ be the origin of rectangular coordinate (X-Y) system.

Let $M(x', y')$ be a point on any of the four quadrant of the X-Y system, then $\overline{OM}$ is the position vector.

Definition 2.4 Let $\tilde{A} = \left(A^{(p)}, A, A^{(o)}\right)$ and $\tilde{B} = \left(B^{(p)}, B, B^{(o)}\right)$ be two triangular fuzzy numbers, then

$$\left(A^{(p)}, A, A^{(o)}\right) \oplus \left(B^{(p)}, B, B^{(o)}\right) \approx \left(A^{(p)} + B^{(p)}, A + B, A^{(o)} + B^{(o)}\right)$$

$$\left(A^{(p)}, A, A^{(o)}\right) \ominus \left(B^{(p)}, B, B^{(o)}\right) \approx \left(A^{(p)} - B^{(o)}, A - B, A^{(o)} - B^{(p)}\right)$$

$$k\left(A^{(p)}, A, A^{(o)}\right) \approx \left(kA^{(p)}, kA, kA^{(o)}\right), k \geq 0,$$

$$k\left(A^{(p)}, A, A^{(o)}\right) \approx \left(kA^o, kA, kA^{(p)}\right), k \leq 0$$

$$\left(A^{(p)}, A, A^{(o)}\right) \otimes \left(B^{(p)}, B, B^{(o)}\right) \approx (A^{(p)}B^{(p)}, AB, A^{(o)}B^{(o)}), \text{ if}$$
$$A^{(p)} \geq 0 \text{ and } B^{(p)} \geq 0$$

$$\left(A^{(p)}, A, A^{(o)}\right) \otimes \left(B^{(p)}, B, B^{(o)}\right) \approx (A^{(p)}B^{(o)}, AB, A^{(o)}B^{(o)}), \text{ if}$$
$$A^{(p)} < 0, A^{(o)} \geq 0, B^{(p)} \geq 0$$

$$\left(A^{(p)}, A, A^{(o)}\right) \otimes \left(B^{(p)}, B, B^{(o)}\right) \approx \left(A^{(p)}B^{(o)}, AB, A^{(p)}B^{(o)}\right), \text{ if}$$
$$A^{(o)} \leq 0 \text{ and } B^{(p)} \geq 0$$

The association between triangular fuzzy number $\tilde{A} = \left(A^p, A, A^o\right)$ and incenter point (x', y') of the triangle is shown in Figure 4.1.

4.3 MULTI-OBJECTIVE FUZZY LINEAR PROGRAMMING PROBLEM

The mathematical model for multi-objective fuzzy linear programming problem with fuzzy decision variable is expressed as:

$$\widetilde{\max} : \tilde{Z}_k \approx \tilde{C}_{k1}\tilde{X}_1 \oplus \tilde{C}_{k2}\tilde{X}_2 \oplus \tilde{C}_{k3}\tilde{X}_3 \ldots \oplus \tilde{C}_{kn}\tilde{X}_n, k = 1, 2, 3, \ldots K \qquad (4.1)$$

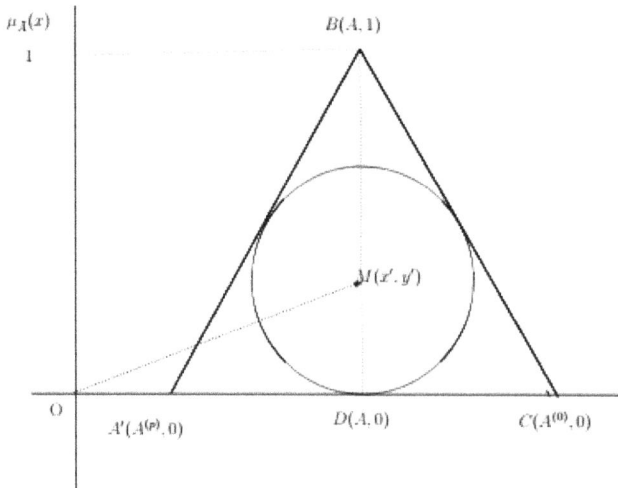

Figure 4.1 The association between triangular fuzzy number.

subject to

$$\tilde{A}_{11}\tilde{X}_1 \oplus \tilde{A}_{12}\tilde{X}_2 \oplus \cdots \oplus \tilde{A}_{1n}\tilde{X}_n \circ \tilde{B}_1 \qquad (4.2)$$

$$\vdots \quad \vdots \qquad\qquad (4.3)$$

$$\tilde{A}_{m1}\tilde{X}_1 \oplus \tilde{A}_{m2}\tilde{X}_2 \oplus \cdots \oplus \tilde{A}_{mn}\tilde{X}_n \circ \tilde{B}_m \qquad (4.4)$$

$$\tilde{X}_j \pm 0, j = 1,2,...,n \qquad (4.5)$$

where $\tilde{C}_{kj} = \left(C_{kj}^{(p)}, C_{kj}, C_{kj}^{(0)} \right)$ is cost-coefficient which is assumed to be triangular fuzzy number,

$\tilde{A}_{ij} = \left(A_{ij}^{(p)}, A_{ij}, A_{ij}^{(0)} \right)$ is technological coefficient which is assumed to be triangular fuzzy number,

$\tilde{B}_i = \left(B_i^{(p)}, B_i, B_i^{(0)} \right)$ is right-hand-side value of the ith constraint which is assumed to be triangular fuzzy number, and

$\tilde{X}_j = \left(X_j^{(p)}, X_j, X_j^{(o)} \right)$ is decision variable which is also assumed to be triangular fuzzy variable.

Note: $\tilde{C}_{kj}\tilde{X}_j = \tilde{C}_{kj} \otimes \tilde{X}_j$ and $\tilde{A}_{ij}\tilde{X}_j = \tilde{A}_{ij} \otimes \tilde{X}_j$

4.4 SOLUTION PROCEDURES

In this section, step-by-step solution procedure is presented to find the optimal solution for multi-objective FLP problem with fuzzy decision variables. The steps of the proposed algorithm are given as follows:

Step 1: Let $\omega_1, \omega_2, \omega_3, \omega_4, \cdots, \omega_k$ be the weights for first, second, third, fourth, . . ., kth objective functions. Employing the weighted sum method, the single-objective FLP problem is expressed as:

$$\widetilde{\max} : \tilde{Z} = \sum_{k=1}^{K} \omega_k \widetilde{Z}_k \qquad (4.6)$$

subject to

$$\tilde{A}_{11}\tilde{X}_1 \oplus \tilde{A}_{12}\tilde{X}_2 \oplus \cdots \oplus \tilde{A}_{1n}\tilde{X}_n \circ \tilde{B}_1 \qquad (4.7)$$

$$\vdots \quad \vdots \qquad\qquad (4.8)$$

$$\tilde{A}_{m1}\tilde{X}_1 \oplus \tilde{A}_{m2}\tilde{X}_2 \oplus \cdots \oplus \tilde{A}_{mn}\tilde{X}_n \circ \tilde{B}_m \qquad 4.9)$$

$$\tilde{X}_j \pm 0 \qquad (4.10)$$

$$\omega_1 + \omega_2 + \omega_3 + \omega_4 + \cdots + \omega_k = 1 \qquad (4.11)$$

Step 2: Consider j-th term of k-th objective function of the proposed mathematical model as:

$$\omega_k\left(C_{kj}^p, C_{kj}, C_{kj}^o\right) \otimes \left(X_j^p, X_j, X_j^o\right) \tag{4.12}$$

and the i-th constraint as:

$$\left(A_{ij}^p, A_{ij}, A_{ij}^o\right) \otimes \left(X_j^p, X_j, X_j^o\right) \circ \left(B_i^p, B_i, B_i^0\right) \tag{4.13}$$

Step 3: Find the incenters for each fuzzy parameters $\left(A_{ij}^p, A_{ij}, A_{ij}^o\right)$, $\left(\overset{\varnothing p}{C_{kj}}, C_{kj}, C_{kj}^o\right)$, and $\left(B_i^p, B_i, B_i^0\right)$ as $\left(\dfrac{a_{ij}^p A_{ij}^p + a_{ij} A_{ij} + a_{ij}^o A_{ij}^o}{P_{ij}}, \dfrac{a_{ij}}{P_{ij}}\right)$, $\left(\dfrac{c_{kj}^p C_{kj}^p + c_{kj} C_{kj} + c_{kj}^o C_{kj}^o}{P_{kj}}, \dfrac{c_{kj}}{P_{kj}}\right)$, and $\left(\dfrac{b_i^p B_i^p + b_i B_i + b_i^o B_i^o}{P_i}, \dfrac{b_i}{P_i}\right)$, respectively.

Also obtain the incenters of each fuzzy variable $\left(X_j^p, X_j, X_j^o\right)$ as $\left(\dfrac{x_j^p X_j^p + x_j X_j + x_j^o X_j^o}{P_{x_j}}, \dfrac{x_j}{P_{x_j}}\right)$ are perimeters of the triangles

where $P_{ij} = a_{ij}^p + a_{ij} + a_{ij}^o$, $P_{kj} = c_{kj}^p + c_{kj} + c_{kj}^o$, $P_i = b_i^p + b_i + b_i^o$, and $P_{x_j} = x_j^p + x_j + x_j^o$.

Step 4: Evaluate the length of position vector associated with each incenters:

The length of position vector associated with cost coefficients is

$$\sqrt{\left(\dfrac{c_{kj}^p C_{kj}^p + c_{kj} C_{kj} + c_{kj}^o C_{kj}^o}{P_{kj}}\right)^2 + \left(\dfrac{c_{kj}}{P_{kj}}\right)^2}.$$

The length of position vector associated with decision variable is

$$\sqrt{\left(\dfrac{x_j^p X_j^p + x_j X_j + x_j^o X_j^o}{P_{x_j}}\right)^2 + \left(\dfrac{x_j}{P_{x_j}}\right)^2}.$$

The length of position vector associated with tech. coefficient is

$$\sqrt{\left(\dfrac{a_{ij}^p A_{ij}^p + a_{ij} A_{ij} + a_{ij}^o A_{ij}^o}{P_{ij}}\right)^2 + \left(\dfrac{a_{ij}}{P_{ij}}\right)^2}.$$

The length of position vector associated with RHS constraint is

$$\sqrt{\left(\dfrac{b_i^p B_i^p + b_i B_i + b_i^o B_i^o}{P_i}\right)^2 + \left(\dfrac{b_i}{P_i}\right)^2}.$$

Step 5: The equivalent crisp non-linear programming problem is given by:

$$\max: Z = \sum_{k=1}^{K} \omega_k \left(\sum_{j=1}^{n} \left(\sqrt{\left(\frac{c_{kj}^p C_{kj}^p + c_{kj} C_{kj} + c_{kj}^o C_{kj}^o}{P_{kj}} \right)^2 + \left(\frac{c_{kj}}{P_{kj}} \right)^2} \times \sqrt{\left(\frac{x_j^p X_j^p + x_j X_j + x_j^o X_j^o}{P_{x_j}} \right)^2 + \left(\frac{x_j}{P_{x_j}} \right)^2} \right) \right)$$
(4.14)

subject to

$$\sum_{j=1}^{n} \left(\sqrt{\left(\frac{a_{ij}^p A_{ij}^p + a_{ij} A_{ij} + a_{ij}^o A_{ij}^o}{P_{ij}} \right)^2 + \left(\frac{a_{ij}}{P_{ij}} \right)^2} \times \sqrt{\left(\frac{x_j^p X_j^p + x_j X_j + x_j^o X_j^o}{P_{x_j}} \right)^2 + \left(\frac{x_j}{P_{x_j}} \right)^2} \right)$$

$$\leq \sqrt{\left(\frac{b_i^p B_i^p + b_i B_i + b_i^o B_i^o}{P_i} \right)^2 + \left(\frac{b_i}{P_i} \right)^2}, i = 1,2,3,..,m$$
(4.15)

$$X_j - X_j^p \geq 0, X_j^o - X_j \geq 0, X_j^p \geq 0, j = 1,2,3,..,n$$
(4.16)

Step 6: The above crisp NLP mathematical model is solved by LINGO Schrage and LINDO Systems (1997) software and a fuzzy compromise solutions, $\tilde{X}_j = \left(X_j^p, X_j, X_j^o \right), j = 1,2,3...,n$ are obtained.

Step 7: For $\tilde{Z}_k, k = 1,2,3,...,K$, the corresponding fuzzy values are calculated by substituting all the above fuzzy PO solutions in each fuzzy objective function given in the model.

4.5 NUMERICAL EXAMPLE

$$\widetilde{\max}: \tilde{Z}_1 \approx \left(1,2,3 \right) \otimes \tilde{X}_1 \oplus \left(2,3,4 \right) \otimes \tilde{X}_2 \oplus \left(3,4,5 \right) X_3$$
(4.17)

$$\widetilde{\max}: \tilde{Z}_2 \approx \left(1,2,6 \right) \otimes \tilde{X}_1 \oplus \left(2,3,7 \right) \otimes \tilde{X}_2 \oplus \left(2,4,8 \right) X_3$$
(4.18)

$$\widetilde{\max}: \tilde{Z}_3 \approx \left(2,4,6 \right) \otimes \tilde{X}_1 \oplus \left(3,5,7 \right) \otimes \tilde{X}_2 \oplus \left(5,7,9 \right) X_3$$
(4.19)

subject to

$$\left(0,1,2 \right) \otimes \tilde{X}_1 \oplus \left(1,2,3 \right) \otimes \tilde{X}_2 \oplus \left(2,3,4 \right) \tilde{X}_3 \circ \left(2,20,54 \right)$$
(4.20)

$$\left(2,3,4 \right) \otimes \tilde{X}_1 \oplus \left(0,1,2 \right) \otimes \tilde{X}_2 \oplus \left(1,2,3 \right) \tilde{X}_3 \circ \left(3,21,55 \right)$$
(4.21)

$$\left(1,2,3 \right) \otimes \tilde{X}_1 \oplus \left(2,3,4 \right) \otimes \tilde{X}_2 \oplus \left(0,1,2 \right) \tilde{X}_3 \circ \left(4,22,56 \right)$$
(4.22)

$$\tilde{X}_1 \pm 0, \tilde{X}_2 \pm 0, \tilde{X}_3 \pm 0 \tag{4.23}$$

Let ω_1, ω_2, and ω_3 be the weights for first, second, and third objective functions. Using the weighted sum method, the single-objective FLP problem is expressed as:

$$\widetilde{\max} : \tilde{Z}^* \approx \omega_1 \tilde{Z}_1 + \omega_2 \tilde{Z}_2 + \omega_3 \tilde{Z}_3 \tag{4.24}$$

subject to

$$(0,1,2) \otimes \tilde{X}_1 \oplus (1,2,3) \otimes \tilde{X}_2 \oplus (2,3,4) \tilde{X}_3 \circ (2,20,54) \tag{4.25}$$

$$(2,3,4) \otimes \tilde{X}_1 \oplus (0,1,2) \otimes \tilde{X}_2 \oplus (1,2,3) \tilde{X}_3 \circ (3,21,55) \tag{4.26}$$

$$(1,2,3) \otimes \tilde{X}_1 \oplus (2,3,4) \otimes \tilde{X}_2 \oplus (0,1,2) \tilde{X}_3 \circ (4,22,56) \tag{4.27}$$

$$\omega_1 + \omega_2 + \omega_3 = 1 \tag{4.28}$$

$$\tilde{X}_1 \pm 0, \tilde{X}_2 \pm 0, \tilde{X}_3 \pm 0 \tag{4.29}$$

Decision-maker has decided the following weights (three cases) by observing the importance of each objective functions as in Table 4.1.

Now the objective functions are expressed as:

$$\widetilde{Z}_1^* \approx (1.4,2.4,4.5) \otimes \tilde{X}_1 \oplus (1.9,3.4,5.5) \otimes \tilde{X}_2 \oplus (3.1,4.6,6.7) \tilde{X}_3 \tag{4.30}$$

$$\widetilde{Z}_2^* \approx (1.3,2.6,6.4) \otimes \tilde{X}_1 \oplus (1.8,3.8,6.4) \otimes \tilde{X}_2 \oplus (3.1,4.9,7.7) \tilde{X}_3 \tag{4.31}$$

$$\widetilde{Z}_3^* \approx (1.5,3.3,5.1) \otimes \tilde{X}_1 \oplus (2.3,3.8,6.1) \otimes \tilde{X}_2 \oplus (3.8,5.5,7.6) \tilde{X}_3 \tag{4.32}$$

By using solution procedure, the defuzzification of fuzzy parameters is demonstrated in Table 4.2.

Table 4.1 Weighted Objectives

Case	ω_1	ω_2	ω_3	$\tilde{Z}^*$
I	0.5	0.3	0.2	$\widetilde{Z}_1^* \approx 0.5\tilde{Z}_1 + 0.3\tilde{Z}_2 + 0.2\tilde{Z}_3$
II	0.2	0.5	0.3	$\widetilde{Z}_2^* \approx 0.2\tilde{Z}_1 + 0.5\tilde{Z}_2 + 0.3\tilde{Z}_3$
III	0.3	0.2	0.5	$\widetilde{Z}_3^* \approx 0.3\tilde{Z}_1 + 0.2\tilde{Z}_2 + 0.5\tilde{Z}_3$

Table 4.2 Defuzzified Form of Fuzzy Parameters

Fuzzy parameter	Defuzzification
Parameters in $\widetilde{Z}_1^*$	$(1.4, 2.4, 4.5) = 2.541, (1.9, 3.4, 5.5) = 3.466, (3.1, 4.6, 6.7) = 4.662$
Parameters in $\widetilde{Z}_2^*$	$(1.3, 2.6, 5.4) = 2.724, (1.8, 3.6, 6.4) = 3.674, (3.1, 4.9, 7.7) = 4.966$
Parameters in $\widetilde{Z}_3^*$	$(1.5, 3.3, 5.1) = 3.330, (2.3, 3.8, 6.1) = 3.880, (3.8, 5.5, 7.6) = 5.543$
Parameters in constraint	$(0, 1, 2) = 1.082, (1, 2, 3) = 2.043, (2, 3, 4) = 3.082$
Parameters in right-hand sides	$(2, 20, 54) = 20.020, (3, 21, 55) = 21.013, (4, 22, 56) = 22.071$

From equation (5.8) and Table 4.2, the equivalent-crisp single-objective NLP problem is given for each case as follows:

Case I:

$$\max: Z = 2.541\sqrt{\left(\frac{x_1^{(p)}X_1^{(p)} + x_1X_1 + x_1^{(o)}X_1^{(o)}}{P_{x1}}\right)^2 + \left(\frac{x_1}{P_{x1}}\right)^2}$$

$$+3.466\sqrt{\left(\frac{x_2^{(p)}X_2^{(p)} + x_2X_2 + x_2^{(o)}X_2^{(o)}}{P_{x2}}\right)^2 + \left(\frac{x_2}{P_{x2}}\right)^2}$$

$$+4.662\sqrt{\left(\frac{x_3^{(p)}X_3^{(p)} + x_3X_3 + x_3^{(o)}X_3^{(o)}}{P_{x3}}\right)^2 + \left(\frac{x_3}{P_{x3}}\right)^2} \qquad (4.33)$$

subject to

$$1.082\sqrt{\left(\frac{x_1^{(p)}X_1^{(p)} + x_1X_1 + x_1^{(o)}X_1^{(o)}}{P_{x1}}\right)^2 + \left(\frac{x_1}{P_{x1}}\right)^2}$$

$$+2.043\sqrt{\left(\frac{x_2^{(p)}X_2^{(p)} + x_2X_2 + x_2^{(o)}X_2^{(o)}}{P_{x2}}\right)^2 + \left(\frac{x_2}{P_{x2}}\right)^2}$$

$$+3.082\sqrt{\left(\frac{x_3^{(p)}X_3^{(p)} + x_3X_3 + x_3^{(o)}X_3^{(o)}}{P_{x3}}\right)^2 + \left(\frac{x_3}{P_{x3}}\right)^2} \leq 20.020 \qquad (4.34)$$

$$3.082\sqrt{\left(\frac{x_1^{(p)}X_1^{(p)} + x_iX_1 + x_1^{(o)}X_1^{(o)}}{P_{x1}}\right)^2 + \left(\frac{x_1}{P_{x1}}\right)^2}$$

$$+1.082\sqrt{\left(\frac{x_2^{(p)}X_2^{(p)} + x_2X_2 + x_2^{(o)}X_2^{(o)}}{P_{x2}}\right)^2 + \left(\frac{x_2}{P_{x2}}\right)^2}$$

$$+2.043\sqrt{(\frac{x_3^{(p)}X_3^{(p)} + x_3X_3 + x^{(o)})_3X_3^{(o)}}{P_{x3}})^2 + (\frac{x_3}{P_{x3}})^2} \leq 21.013 \qquad (4.35)$$

$$+2.043\sqrt{(\frac{x_1^{(p)}X_1^{(p)} + x_jX_1 + x_1^{(o)}X_1^{(o)}}{P_{x1}})^2 + (\frac{x_1}{P_{x1}})^2}$$

$$+3.082\sqrt{(\frac{x_2^{(p)}X_2^{(p)} + x_2X_2 + x_2^{(o)}X_2^{(o)}}{P_{x2}})^2 + (\frac{x_2}{P_{x2}})^2}$$

$$1.082\sqrt{(\frac{x_3^{(p)}X_3^{(p)} + x_3X_3 + x_3^{(o)}X_3^{(o)}}{P_{x3}})^2 + (\frac{x_3}{P_{x3}})^2} \leq 22.071 \qquad (4.36)$$

$$P_{xj} = x_j^{(p)} + x_j + x_j^{(o)} \qquad (4.37)$$

$$x_j^{(p)} = \sqrt{(X_j - X_j^{(o)})^2 + 1} \qquad (4.38)$$

$$x_j = X_j^{(o)} - X_j^{(p)} \qquad (4.39)$$

$$x_j^{(o)} = \sqrt{(X_j^{(p)} - X_1)^2 + 1} \qquad (4.40)$$

$$x_j^{(P)} \geq 0, x_j \geq 0, x_j^{(o)} \geq 0, j = 1, 2, 3 \qquad (4.41)$$

Case II:

$$\max : Z = 2.724\sqrt{(\frac{x_1^{(p)}X_1^{(p)} + x_1X_1 + x_1^{(o)}X_1^{(o)}}{P_{x1}})^2 + (\frac{x_1}{P_{x1}})^2}$$

$$+3.674\sqrt{(\frac{x_2^{(p)}X_2^{(p)} + x_2X_2 + x_2^{(o)}X_2^{(o)}}{P_{x2}})^2 + (\frac{x_2}{P_{x2}})^2}$$

$$+5.543\sqrt{(\frac{x_3^{(p)}X_3^{(p)} + x_3X_3 + x_3^{(o)}X_3^{(o)}}{P_{x3}})^2 + (\frac{x_3}{P_{x3}})^2} \qquad (4.42)$$

subject to

$$1.082\sqrt{(\frac{x_1^{(p)}X_1^{(p)} + x_1X_1 + x_1^{(o)}X_1^{(o)}}{P_{x1}})^2 + (\frac{x_1}{P_{x1}})^2}$$

$$+2.043\sqrt{(\frac{x_2^{(p)}X_2^{(p)} + x_2X_2 + x_2^{(o)}X_2^{(o)}}{P_{x2}})^2 + (\frac{x_2}{P_{x2}})^2}$$

$$+3.082\sqrt{(\frac{x_3^{(p)}X_3^{(p)} + x_3X_3 + x_3^{(o)}X_3^{(o)}}{P_{x3}})^2 + (\frac{x_3}{P_{x3}})^2} \leq 20.020 \qquad (4.43)$$

$$3.082\sqrt{(\frac{x_1^{(p)}X_1^{(p)} + x_jX_1 + x_1^{(o)}X_1^{(o)}}{P_{x1}})^2 + (\frac{x_1}{P_{x1}})^2}$$

$$+1.082\sqrt{(\frac{x_2^{(p)}X_2^{(p)} + x_2X_2 + x_2^{(o)}X_2^{(o)}}{P_{x2}})^2 + (\frac{x_2}{P_{x2}})^2}$$

$$+2.043\sqrt{(\frac{x_3^{(p)}X_3^{(p)} + x_3X_3 + x^{(o)}{}_3X_3^{(o)}}{P_{x3}})^2 + (\frac{x_3}{P_{x3}})^2} \le 21.013 \tag{4.44}$$

$$+2.043\sqrt{(\frac{x_1^{(p)}X_1^{(p)} + x_jX_1 + x_1^{(o)}X_1^{(o)}}{P_{x1}})^2 + (\frac{x_1}{P_{x1}})^2}$$

$$+3.082\sqrt{(\frac{x_2^{(p)}X_2^{(p)} + x_2X_2 + x_2^{(o)}X_2^{(o)}}{P_{x2}})^2 + (\frac{x_2}{P_{x2}})^2}$$

$$1.082\sqrt{(\frac{x_3^{(p)}X_3^{(p)} + x_3X_3 + x_3^{(o)}X_3^{(o)}}{P_{x3}})^2 + (\frac{x_3}{P_{x3}})^2} \le 22.071 \tag{4.45}$$

$$P_{xj} = x_j^{(p)} + x_j + x_j^{(o)} \tag{4.46}$$

$$x_j^{(p)} = \sqrt{(X_j - X_j^{(o)})^2 + 1} \tag{4.47}$$

$$x_j = X_j^{(o)} - X_j^{(p)} \tag{4.48}$$

$$x_j^{(o)} = \sqrt{(X_j^{(p)} - X_1)^2 + 1} \tag{4.49}$$

$$x_j^{(p)} \ge 0, x_j \ge 0, x_j^{(o)} \ge 0, j = 1,2,3 \tag{4.50}$$

Case III:

$$\max : Z = \frac{3.350}{P_{x1}} \sqrt{(\frac{x_1^{(p)}X_1^{(p)} + x_1X_1 + x_1^{(o)}X_1^{(o)}}{P_{x1}})^2 + (\frac{x_1}{P_{x1}})^2}$$

$$+\frac{3.880}{P_{x2}} \sqrt{(\frac{x_2^{(p)}X_2^{(p)} + x_2X_2 + x_2^{(o)}X_2^{(o)}}{P_{x2}})^2 + (\frac{x_2}{P_{x2}})^2}$$

$$+\frac{5.543}{P_{x3}} \sqrt{(\frac{x_3^{(p)}X_3^{(p)} + x_3X_3 + x_3^{(o)}X_3^{(o)}}{P_{x3}})^2 + (\frac{x_3}{P_{x3}})^2} \tag{4.51}$$

subject to

$$1.082\sqrt{(\frac{x_1^{(p)}X_1^{(p)} + x_1X_1 + x_1^{(o)}X_1^{(o)}}{P_{x1}})^2 + (\frac{x_1}{P_{x1}})^2}$$

$$+2.043\sqrt{(\frac{x_2^{(p)}X_2^{(p)} + x_2X_2 + x_2^{(o)}X_2^{(o)}}{P_{x2}})^2 + (\frac{x_2}{P_{x2}})^2}$$

$$+3.082\sqrt{(\frac{x_3^{(p)}X_3^{(p)} + x_3X_3 + x_3^{(o)}X_2^{(o)}}{P_{x3}})^2 + (\frac{x_3}{P_{x3}})^2} \le 20.020 \qquad (4.52)$$

$$3.082\sqrt{(\frac{x_1^{(p)}X_1^{(p)} + x_jX_1 + x_1^{(o)}X_1^{(o)}}{P_{x1}})^2 + (\frac{x_1}{P_{x1}})^2}$$

$$+1.082\sqrt{(\frac{x_2^{(p)}X_2^{(p)} + x_2X_2 + x_2^{(o)}X_2^{(o)}}{P_{x2}})^2 + (\frac{x_2}{P_{x2}})^2}$$

$$+2.043\sqrt{(\frac{x_3^{(p)}X_3^{(p)} + x_3X_3 + x^{(o)}{}_3X_3^{(o)}}{P_{x3}})^2 + (\frac{x_3}{P_{x3}})^2} \le 21.013 \qquad (4.53)$$

$$2.043\sqrt{(\frac{x_1^{(p)}X_1^{(p)} + x_jX_1 + x_1^{(o)}X_1^{(o)}}{P_{x1}})^2 + (\frac{x_1}{P_{x1}})^2}$$

$$+3.082\sqrt{(\frac{x_2^{(p)}X_2^{(p)} + x_2X_2 + x_2^{(o)}X_2^{(o)}}{P_{x2}})^2 + (\frac{x_2}{P_{x2}})^2}$$

$$+1.082\sqrt{(\frac{x_3^{(p)}X_3^{(p)} + x_3X_3 + x_3^{(o)}X_3^{(o)}}{P_{x3}})^2 + (\frac{x_3}{P_{x3}})^2} \le 22.071 \qquad (4.54)$$

$$P_{xj} = x_j^{(p)} + x_j + x_j^{(o)} \qquad (4.55)$$

$$x_j^{(p)} = \sqrt{(X_j - X_j^{(o)})^2 + 1} \qquad (4.56)$$

$$x_j = X_j^{(o)} - X_j^{(p)} \qquad (4.57)$$

$$x_j^{(o)} = \sqrt{(X_j^{(p)} - X_1)^2 + 1} \qquad (4.58)$$

$$x_j^{(p)} \ge 0, x_j \ge 0, x_j^{(o)} \ge 0, j = 1,2,3 \qquad (4.59)$$

Solving all the above cases according to their order of importance, the following fuzzy PO solutions are obtained.

Case 1: $\tilde{X}_1 = (2.964367, 4, 4.041728), \tilde{X}_2 = (3, 3, 4.424189)$, and
$\tilde{X}_3 = (2.821132, 3, 3.068088)$

Case 2: $\tilde{X}_1 = (2.386766, 4, 4.093556), \tilde{X}_2 = (2.92387, 4, 4.032624)$, and
$\tilde{X}_3 = (1.969680, 3, 3)$

Case 3: $\tilde{X}_1 = (2.66748, 4, 4.047266), \tilde{X}_2 = (3.090336, 4, 4)$, and
$\tilde{X}_3 = (1.947234, 3, 3.005338)$

The PO solutions are substituted in equation (3.1) and fuzzy optimal values for each case are obtained and given in Table 4.3.

4.6 CONCLUSION

The proposed method is easy and more accurate for dealing with real-life problem with multi-objectives under uncertainty and vagueness. We have considered solution of multi-objective FLP problem with fuzzy decision variables using the incenter of a triangle procedure. Fuzzy optimal solutions are obtained in reference of integer target values (core values), that is, $x_j \geq 0$ and x_j integer for $j = 1, 2, 3, \dots n$ and fuzzy optimal values are evaluated further at values of decision variables. To explain the proposed system, various numerical examples are solved and outcomes are meticulously checked. To reach on fuzzy solution, we first convert all the fuzziness of parameter and fuzzy decision variables into their corresponding single number equivalence, then the resultant crisp LP problem is solved by LINGO software.

Moreover, our proposed computational algorithm may be implemented for further study in the areas such as various real-life problems with multiple objectives under uncertain environment, decision-making of real-life multiple-objective transportation, and assignment problems.

Table 4.3 Fuzzy Functional Values

Case	$\widetilde{Z}_1$	$\widetilde{Z}_2$	$\widetilde{Z}_3$
I	(14.590684, 32, 43.20156)	(9.05125, 32, 76.40302)	(24.095836, 57, 87.53736)
II	(14.6036102, 32, 43.194859)	(9.5180958, 32, 76.389718)	(24.121706, 57, 79.389718)
III	(14.111776, 32, 43.477888)	(9.094036, 32, 77.239324)	(17.293976, 48, 77.239324)

REFERENCES

Acharya, S. & Biswal, M. P. (2015), 'Application of multi-choice fuzzy linear programming problem to a garment manufacture company', *Journal of Information and Optimization Sciences* 36(6), 569–593.

Allahviranloo, T., Lotfi, F. H., Kiasary, M. K., Kiani, N. & Alizadeh, L. (2008), 'Solving fully fuzzy linear programming problem by the ranking function', *Applied Mathematical Sciences* 2(1), 19–32.

Bellman, R. E. & Zadeh, L. A. (1970), 'Decision-making in a fuzzy environment', *Management Science* 17(4), B-141.

Buckley, J. J. (1988), 'Possibilistic linear programming with triangular fuzzy numbers', *Fuzzy Set and Systems* 26(1), 135–138.

Buckley, J. J. (1989), 'Solving possibilistic linear programming problems', *Fuzzy Sets and Systems* 31(3), 329–341.

Buckley, J. J. & Feuring, T. (2000), 'Evolutionary algorithm solution to fuzzy problems: Fuzzy linear programming', *Fuzzy Sets and Systems* 109(1), 35–53.

Das, S. (2017), 'Modified method for solving fully fuzzy linear programming problem with triangular fuzzy numbers', *International Journal of Research in Industrial Engineering* 6(4), 293–311.

Deshmukh, M. C., Ghadle, K. P. & Jadhav, O. S. (2020), 'Optimal solution of fully fuzzy lPP with symmetric HFNS', *in Computing in Engineering and Technology*, Springer, pp. 387–395.

Hamadameen, A. O. & Hassan, N. (2018), 'A compromise solution for the fully fuzzy multi-objective linear programming problems', *IEEE Access* 6, 43696–43711.

Hashemi, S. M., Modarres, M., Nasrabadi, E. & Nasrabadi, M. M. (2006), 'Fully fuzzified linear programming, solution and duality', *Journal of Intelligent & Fuzzy Systems* 17(3), 253–261.

Inuiguchi, M., Ichihashi, H. & Tanaka, H. (1990), 'Fuzzy programming: A survey of recent developments', *in Stochastic Versus Fuzzy Approaches to Multi-Objective Mathematical Programming Under Uncertainty*, Springer, pp. 45–68.

Jalil, S. A., Sadia, S., Javaid, S. & Ali, Q. (2017), 'A solution approach for solving fully fuzzy multi-objective solid transportation problem', *International Journal of Agricultural and Statistical Sciences* 13(1), 75–84.

Julien, B. (1994), 'An extension to possibilistic linear programming', *Fuzzy Sets and Systems* 64(2), 195–206.

Khalifa, H. A. (2019), 'Utilizing a new approach for solving fully fuzzy linear programming problems', *Croatian Operational Research Review* 10(2), 337–344.

Kumar, A., Kaur, J. & Singh, P. (2011), 'A new method for solving fully fuzzy linear programming problems', *Applied Mathematical Modelling* 35(2), 817–823.

Maleki, H. R. (2002), 'Ranking functions and their applications to fuzzy linear programming', *Far East Journal of Mathematical Sciences* 4(3), 283–302.

Mitlif, R. J. (2019), 'A new ranking function of triangular fuzzy numbers for solving fuzzy linear programming problems with big-m method', *Electronics Science Technology and Application* 6(2), 10–13.

Nasseri, S., Behmanesh, E., Taleshian, F., Abdolalipoor, M. & Taghi, N. N. (2013), 'Fully fuzzy linear programming with inequality constraints', *International Journal of Industrial Mathematics* 5(4), 309–316.

Schrage, L. E. & LINDO Systems. (1997), *Optimization modeling with LINGO*, Duxbury Press.

Sharma, U. & Aggarwal, S. (2018), 'Solving fully fuzzy multi-objective linear programming problem using nearest interval approximation of fuzzy number and interval programming', *International Journal of Fuzzy Systems* 20(2), 488–499.

Chapter 5

On approximation of piecewise linear membership functions and its application to solve solid transportation problem with fuzzy cost coefficients

Pradip Kundu and Manidatta Ray

5.1 INTRODUCTION

The solid transportation problem (STP), initially developed by Schell [1] and Haley [2], is an extension of classical transportation problem (TP). Besides source and destination constraints, STP contains an another constraint describing the limitation of the capacities of conveyance (modes of transportation). The volume and complexity of the collected data in various fields including transportation system are growing rapidly. Due to insufficient information, fluctuating financial market, lack of evidence, the available data from previous experiments, or the forecasted values of parameters of a transportation system such as resources, demands, and conveyance capacities are not always crisp or precise but are imprecise. For example, transportation cost depends upon fuel price, labor charges, tax charges, etc., each of which fluctuates time to time. Fuzzy set theory [3, 4, 5, 6] is a widely used efficient tool to deal with impreciseness. There are many methods available in literature to solve optimization problems with fuzzy parameters, for example, chance-constrained programming [7, 8] and expected value model [9] for fuzzy programming. Transportation problems with fuzzy parameters have been studied by many researchers [10, 11, 12, 13, 14, 15, 8, 16, 17, 18, 19, 20]. Also it is observed that trapezoidal and triangular fuzzy numbers are very important and most widely used fuzzy numbers in the literature.

The main problem with the fuzzy numbers like trapezoidal and triangular is that membership functions of these kind of fuzzy numbers are piecewise linear and so are not continuously differentiable. So the gradient-based optimization methods cannot be applied directly to solve the decision-making problem with such kind of fuzzy parameters. So to solve such problems, one has to apply some defuzzification methods (which tends to lose some

DOI: 10.1201/9781003462422-5

information hidden in the uncertain data) either to obtain corresponding crisp form or to apply time-consuming and complex process like fuzzy simulation and evolutionary algorithm. Also most of methods supply the optimal crisp value of the fuzzy objective that does not give a proper idea about objective value according to fuzzy coefficients of the objective function. Dombi and Gera [21] proposed a continuous approximation of piecewise linear membership function using approximation of the cut function. With a general error formula, we show that this approximation can have almost negligible error. The approximated membership functions have continuous gradient and hence can be used to solve optimization problem with fuzzy parameters, to fine tune a fuzzy control system by gradient-based technique, and in neuro-fuzzy learning models, which are mostly based on gradient descent strategies. Copeland et al. [22] used such approximated membership functions so that they can be used in gradient descent methods to optimize an error function. Guimarães et al. [23] discussed the utilization of continuously differentiable membership functions in fuzzy optimization with an application to the optimization of an electrostatic micromotor. In this chapter, we improve the approximation method proposed by Dombi and Gera [21]. The improved method is efficient and gives better approximation of a piecewise linear membership function. We also extend the method to provide approximations of membership functions of some generalized fuzzy numbers.

Next we consider an STP with unit transportation costs as fuzzy numbers. Since the cost parameters are fuzzy numbers and the decision variables are zero or positive real numbers, the objective function also becomes a fuzzy quantity. For illustration here, we take cost parameters as trapezoidal fuzzy numbers. To solve the problem, we first approximate the piecewise linear membership function of the fuzzy objective into a continuously differentiable function using the improved Dombi and Gera's [21] method. Then we reconstruct the problem using this continuous approximation. Now the converted problem can be optimized using gradient-based optimization method, and the optimal parameters of the fuzzy objective function are obtained. Thus for the objective function with fuzzy parameters, we obtain a fuzzy solution (which provide detail information of the objective value) in an elegant way. To provide a comparison to the solution of the problem, we apply an another method using the concept of minimum of fuzzy number [24] which also gives a fuzzy solution for the fuzzy objective function. The problem with coefficients of the objective as triangular fuzzy numbers can be treated in similar way. A numerical experiment has been done to illustrate the two methods and corresponding solutions of the problem are obtained using LINGO solver based upon generalized reduced gradient (GRG) technique.

5.2 FUZZY NUMBER

A fuzzy subset $\tilde{A}$ of real number $\mathfrak{R}$ with membership function $\mu_{\tilde{A}} : \mathfrak{R} \to [0,1]$ is said to be a fuzzy number [25] if

(i) $\mu_{\tilde{A}}(x)$ is upper semi-continuous membership function,

(ii) $\tilde{A}$ is normal, that is, $\exists$ an element x_0 s.t. $\mu_{\tilde{A}}(x_0) = 1$,

(iii) $\tilde{A}$ is fuzzy convex, that is,
$$\mu_{\tilde{A}}(\lambda x_1 + (1-\lambda)x_2) \geq \mu_{\tilde{A}}(x_1) \wedge \mu_{\tilde{A}}(x_2) \forall x_1, x_2 \in \mathfrak{R} \text{ and } \lambda \in [0,1]$$

(iv) Support of $\tilde{A} = \{x \in \mathfrak{R} : \mu_{\tilde{A}}(x) > 0\}$ is bounded.

Trapezoidal fuzzy number (TrFN): A TrFN $\tilde{\xi}$ is a fuzzy number fully determined by quadruplet (u_1, u_2, u_3, u_4) of crisp numbers with $u_1 < u_2 \leq u_3 < u_4$, whose membership function is given by

$$\mu_{\tilde{\xi}}(x) = \begin{cases} \dfrac{x - u_1}{u_2 - u_1}, & \text{if } u_1 \leq x \leq u_2; \\ 1, & \text{if } u_2 \leq x \leq u_3; \\ \dfrac{u_4 - x}{u_4 - u_3}, & \text{if } u_3 \leq x \leq u_4; \\ 0, & \text{otherwise.} \end{cases}$$

$[u_2, u_3]$ is called the core of the TrFN $\tilde{\xi}$ and $(u_2 + u_3)/2$ is the center of the core. If in particular $u_2 = u_3$, then this fuzzy number becomes a triangular fuzzy number (TFN).

Generalized trapezoidal fuzzy number:

A generalized trapezoidal fuzzy number $\tilde{\xi}$ is a trapezoidal fuzzy number which may or may not be normalized. It is determined by $(u_1, u_2, u_3, u_4; w)$ with $0 < w \leq 1$ and its membership function is given by

$$\mu_{\tilde{\xi}}(x) = \begin{cases} \dfrac{w(x - u_1)}{u_2 - u_1}, & \text{if } u_1 \leq x \leq u_2; \\ w, & \text{if } u_2 \leq x \leq u_3; \\ \dfrac{w(u_4 - x)}{u_4 - u_3}, & \text{if } u_3 \leq x \leq u_4; \\ 0, & \text{otherwise.} \end{cases}$$

Here w is called its height, and in particular if $w = 1$, then $\tilde{\xi}$ is usual (i.e., normalized) TrFN.

5.3 CUT FUNCTION AND ITS APPROXIMATION

Definition 1. The cut function (denoted by $[\cdot]$) of the Łukasiewicz (/nilpotent) operator class [26, 27] can be derived from x by taking the maximum of 0 and x and then taking the minimum of the result and 1 (Dombi and Gera [21]), that is,

$$[x] = \min\left(\max(0,x),1\right) = \begin{cases} 0, & \text{if } x \le 0; \\ x, & \text{if } 0 < x < 1; \\ 1, & \text{if } 1 \le x. \end{cases}$$

The generalized cut function is defined as

$$[x]_{a,b} = \left[(x-a)/(b-a)\right] = \begin{cases} 0, & \text{if } x \le a; \\ (x-a)/(b-a), & \text{if } a < x < b; \\ 1, & \text{if } b \le x. \end{cases}$$

where $a,b \in \Re$ and $a < b$.

A nilpotent operator is constructed using the cut function. For example, nilpotent conjunction is represented as $c(x,y) = [x+y-1]$, where $x,y \in [0,1]$.

From the definition, it is seen that the cut function and so that the operators constructed by the cut function do not have continuous gradient. So gradient-based optimization techniques are not applicable with this operator. To overcome this problem, Dombi and Gera [21] introduced a continuously differentiable approximation of the generalized cut function by means of the sigmoid function as defined here.

The sigmoid function is defined as

$$\sigma_d^{(\beta)}(x) = \frac{1}{1 + e^{-\beta(x-d)}}$$

where the lower index d is omitted if 0. They obtained the interval $[a,b]$ squashing function by integrating the difference of two sigmoid functions $\sigma_a^{(\beta)}(x)$ and $\sigma_b^{(\beta)}(x)$ $(a < b)$ as

$$S_{a,b}^{(\beta)}(x) = \frac{1}{b-a} \ln\left(\frac{\sigma_b^{(-\beta)}(x)}{\sigma_a^{(-\beta)}(x)}\right)^{1/\beta}$$

$$= \frac{1}{b-a} \ln\left(\frac{1 + e^{\beta(x-a)}}{1 + e^{\beta(x-b)}}\right)^{1/\beta}, \tag{5.1}$$

where the parameters a and b affect the placement of the interval squashing function.

Dombi and Gera [21] showed that, in the interval $\lfloor a,b \rfloor$, squashing function $S_{a,b}^{(\beta)}(x)$ is an approximation of the generalized cut function $[x]_{a,b}$ by proving the following theorem:

Theorem 3.1 Let $a,b \in \Re$, $a < b$ and $\beta \in \Re^+$. Then

$$\lim_{\beta \to \infty} S_{a,b}^{(\beta)}(x) = [x]_{a,b}$$

and $S_{a,b}^{(\beta)}(x)$ is continuous in x, a, b, and β, where β parameter derives the precision of the approximation.

As an example, the generalized cut function $[x]_{2,4}$ and its approximations $S_{2,4}^{(\beta)}(x)$ for different values of β namely $\beta = 10, 32$ and 38 are shown in Figure 5.1. From the figure, we observe that the generalized cut function and its approximations are almost coincident.

From definition of $S_{a,b}^{(\beta)}(x)$, the following properties are follows:

$$\lim_{\beta \to 0} S_{a,b}^{(\beta)}(x) = 1/2, S_{a,b}^{(-\beta)}(x) = 1 - S_{a,b}^{(\beta)}(x).$$

Also it is obvious that $S_{0,1}^{(\beta)}(x)$ is the approximation of $[x]_{0,1} = [x]$, that is, the cut function.

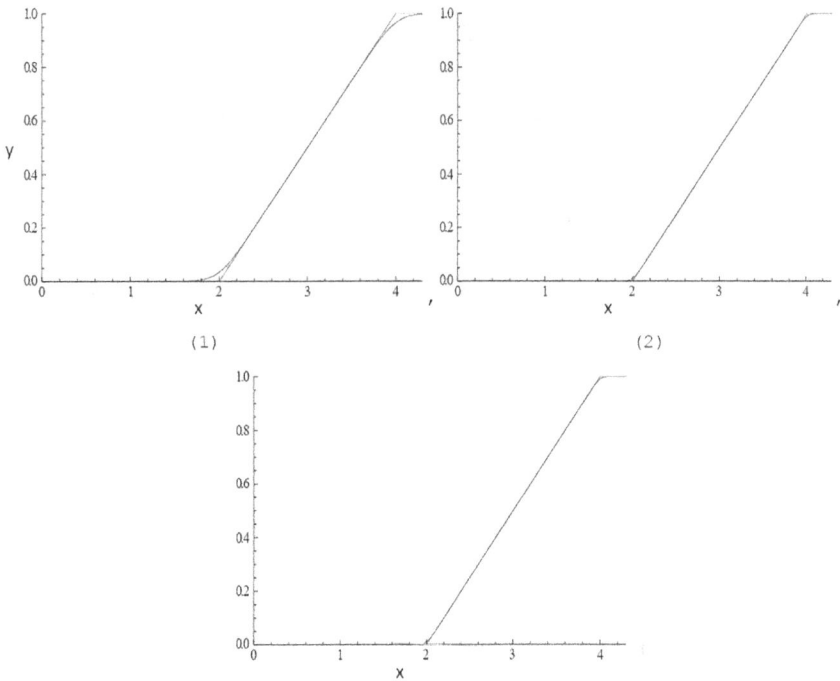

Figure 5.1 $[x]_{2,4}$ and its approximations $S_{2,4}^{(\beta)}(x)$ for (1) $\beta = 10$, (2) $\beta = 32$, and (3) $\beta = 38$.

5.3.1 Error of approximation of $[x]_{a,b}$ by $S_{a,b}^{(\beta)}(x)$ for $\beta > 0$

From Theorem 3.1, a natural question arises that, for a finite value of $\beta(> 0)$, what is the error of approximation of $[x]_{a,b}$ by $S_{a,b}^{(\beta)}(x)$. Here we calculate this error by the following way.

$$S_{a,b}^{(\beta)}(x) = \frac{1}{b-a} \ln(\frac{1+e^{\beta(x-a)}}{1+e^{\beta(x-b)}})^{1/\beta}$$

$$= \frac{1}{b-a} \ln(\frac{e^{\beta(x-a)}(e^{-\beta(x-a)}+1)}{1+e^{\beta(x-b)}})^{1/\beta}$$

$$= \frac{1}{b-a}\left[\ln e^{(x-a)} + \ln(\frac{e^{-\beta(x-a)}+1}{1+e^{\beta(x-b)}})^{1/\beta}\right]$$

$$= \frac{x-a}{b-a} + \frac{1}{b-a} \ln(\frac{e^{-\beta(x-a)}+1}{1+e^{\beta(x-b)}})^{1/\beta}.$$

Now from definition of $[x]_{a,b}$, it is clear that the absolute error of approximation of $[x]_{a,b}$ by $S_{a,b}^{(\beta)}(x)$ for $\beta(> 0)$ is

$$E_{a,b}^{(\beta)}(x) = \frac{1}{b-a} \ln(\frac{e^{-\beta(x-a)}+1}{1+e^{\beta(x-b)}})^{1/\beta}, a \le x \le b, \tag{5.2}$$

which tends to zero as $\beta \to \infty$ and so that the error is almost negligible for large β.

As an example, for the cut function $[x]_{2,4}$, the absolute error of its approximation by $S_{2,4}^{(\beta)}(x)$ for $\beta = 32$ and for a particular value of $x = 2.5$ is $E_{2,4}^{(32)}(x) = 0.1758362 \times 10^{-8}$.

5.4 APPROXIMATION OF PIECEWISE LINEAR MEMBERSHIP FUNCTION

As the cut function can be approximated to a continuously differentiable function, the piecewise linear membership functions representable by the cut function can also be approximated by continuously differentiable functions. In the following, we discuss about approximation of some well-known membership functions.

Approximation of trapezoidal and triangular membership functions: Trapezoidal and triangular fuzzy numbers are the most widely used and important kind of fuzzy numbers. However, their membership functions are piecewise linear and so are not continuously differentiable. Here we discuss approximations of these piecewise linear membership functions to continuously differentiable membership functions.

Consider a trapezoidal fuzzy number (TrFN) $\tilde{A} = (a,b,c,d)$, $a < b \leq c < d$. Dombi and Gera [21] proposed that a trapezoid membership function can be approximated using the conjunction operator and two squashing functions. According to them, the approximation of trapezoid membership function is

$$S_{0,1}^{(\beta)}(S_{a,b}^{(\beta)}(x) - S_{c,d}^{(\beta)}(x))$$

$$= \ln(\frac{1+e^{\beta y}}{1+e^{\beta(y-1)}})^{1/\beta} \tag{5.3}$$

where $y = S_{a,b}^{(\beta)}(x) - S_{c,d}^{(\beta)}(x)$. This result follows from the fact that

$$c([x]_{a,b}, 1-[x]_{c,d}) = \left[[x]_{a,b} - [x]_{c,d}\right],$$

and $S_{a,b}^{(\beta)}(x)$ and $S_{0,1}^{(\beta)}(x)$ are the approximations of $[x]_{a,b}$ and $[x]$, respectively.

However, it is not clear why the nilpotent conjunction is used in the above approximation process. Here, we show that use of nilpotent conjunction is not required and this may increase the error of approximation. We also improve the above approximation process to a more simple process.

A trapezoidal fuzzy membership function can be constructed as difference of two generalized cut functions as

$$[x]_{a,b} - [x]_{a,b} = \begin{cases} 0, & \text{if } x \leq a; \\ \dfrac{x-a}{b-a}, & \text{if } a < x < b; \\ 1, & \text{if } b \leq x \leq c; = \mu_{\tilde{A}}(x), \\ \dfrac{d-x}{d-c}, & \text{if } c < x < d; \\ 0, & \text{if } x \geq d. \end{cases} \tag{5.4}$$

where $\mu_{\tilde{A}}(x)$ is the membership function of TrFN $\tilde{A}$.

Now since $S_{a,b}^{(\beta)}(x)$ is the approximations of $[x]_{a,b}$, from (4) we have the approximation of trapezoidal membership function as

$$\mu_{\tilde{A}}^{app}(x) = S_{a,b}^{(\beta)}(x) - S_{c,d}^{(\beta)}(x). \tag{5.5}$$

Obviously, this approximation is continuously differentiable with respect to x. Also the error the approximation decreases as the value of the parameter β increases.

For example, consider a trapezoidal fuzzy number $\tilde{A} = (2,4,6,8)$. Now with $\beta = 38$, the membership function of $\tilde{A}$ and its approximation are obtained by the improve method (i.e., using equation 5.5), and method of Dombi and Gera [21] is shown in Figure 5.2. From the figure, it is observed that the improve method gives better approximation.

Similarly, for a triangular fuzzy number (TFN) $(a,b,c) \sim (a,b,b,c)$, its membership function can be approximated as

$$S_{a,b}^{(\beta)}(x) - S_{b,c}^{(\beta)}(x).$$

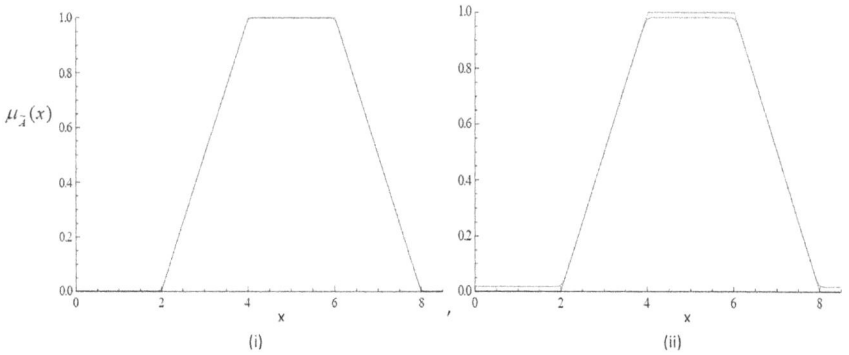

Figure 5.2 TrFN $\tilde{A}$ and its approximation by (i) improve method and (ii) existing method.

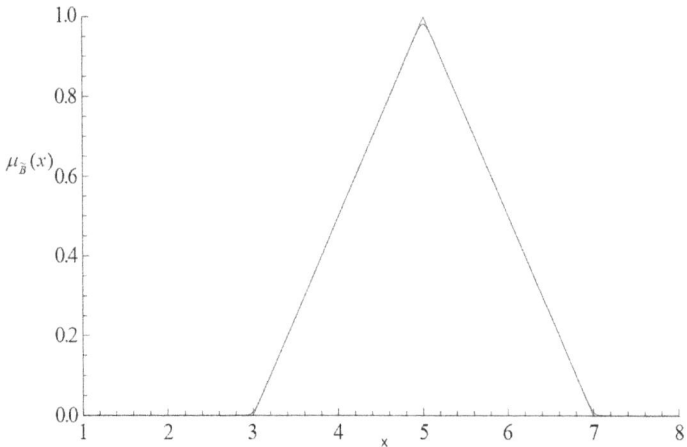

Figure 5.3 Membership function of TFN $\tilde{\beta}$ and its approximation.

For example, consider a triangular fuzzy number $\tilde{B} = (3,5,7)$. In Figure 5.3, the triangular membership function and its approximation for $\beta = 38$ are presented.

In the following, we present an approximation process to make it applicable in case of some generalized fuzzy membership functions.

Approximation of generalized fuzzy membership functions: For approximation of generalized trapezoidal, triangular membership functions, we need to define the following functions.

Definition 2. A cut function of height w $(0 < w \le 1)$, denoted by $[x]^w$, is defined as follows:

$$[x]^w = min\left(max(0,wx),w\right) = \begin{cases} 0, & \text{if } x \le 0; \\ wx, & \text{if } 0 < x < 1; \\ w, & \text{if } 1 \le x. \end{cases}$$

The generalized cut function of height w is defined as

$$[x]^w_{a,b} = \left[(x-a)/(b-a)\right]^w = \begin{cases} 0, & \text{if } x \le a; \\ \dfrac{w(x-a)}{b-a}, & \text{if } a < x < b; \\ w, & \text{if } b \le x. \end{cases}$$

where $a,b \in \Re$ and $a < b$.

In this case, the interval $[a,b]$ squashing function is defined as

$$S^{(\beta)}_{a,b,w}(x) = \frac{w}{b-a}\int(\sigma^{(\beta)}_a(x) - \sigma^{(\beta)}_b(x))dx$$

$$= \frac{w}{b-a}\ln(\frac{1+e^{\beta(x-a)}}{1+e^{\beta(x-b)}})^{1/\beta}. \tag{5.6}$$

We present the following theorem without proof, which shows that $S^{(\beta)}_{a,b,w}(x)$ is an approximation of the generalized cut function $[x]^w_{a,b}$.

Theorem 4.1 *Let* $a,b \in \Re$, $a < b$ $0 < w \le 1$ *and* $\beta \in \Re^+$. *Then*

$$\lim_{\beta \to \infty} S^{(\beta)}_{a,b,w}(x) = [x]^w_{a,b}$$

and $S^{(\beta)}_{a,b,w}(x)$ is continuous in x,a,b, and β.

With the help of the above modified definition of cut function and its approximation, we can now approximate membership functions of some generalized and complicated fuzzy numbers.

Consider a generalized trapezoidal fuzzy number (TrFN) $\tilde{C} = (a,b,c,d;w)$, $a < b \leq c < d$, $0 < w \leq 1$ with membership function as given in Section 5.2. Then generalized trapezoidal fuzzy membership function can be constructed as difference of two generalized cut functions of height w as

$$[x]_{a,b}^{w} - [x]_{c,d}^{w} = \begin{cases} 0, & \text{if } x \leq a; \\ \dfrac{w(x-a)}{b-a}, & \text{if } a < x < b; \\ w, & \text{if } b \leq x \leq c; = \mu_{\tilde{C}}(x), \\ \dfrac{w(d-x)}{d-c}, & \text{if } c < x < d; \\ 0, & \text{if } x \geq d. \end{cases} \qquad (5.7)$$

where $\mu_{\tilde{C}}(x)$ is the membership function of the generalized TrFN $\tilde{C}$.

Now since $S_{a,b,w}^{(\beta)}(x)$ is the approximations of $[x]_{a,b}^{w}$, from (7) we have the approximation of the generalized trapezoidal membership function as

$$\mu_{\tilde{C}}^{app}(x) = S_{a,b,w}^{(\beta)}(x) - S_{c,d,w}^{(\beta)}(x). \qquad (5.8)$$

Obviously this approximation is continuously differentiable with respect to x.

For example, consider a generalized trapezoidal fuzzy number $\tilde{C} = (2,4,6,8;0.8)$. Now with $\beta = 38$, the membership function of $\tilde{C}$ and its approximation are shown in Figure 5.4.

Approximation of membership function of α_0 -piecewise linear 1-knot fuzzy number: Membership function of an α_0 -piecewise linear 1-knot fuzzy number [28]) $\tilde{D} = \tilde{D}(\alpha_0, (a_1, a_2, a_3, a_4, a_5, a_6))$ is given by

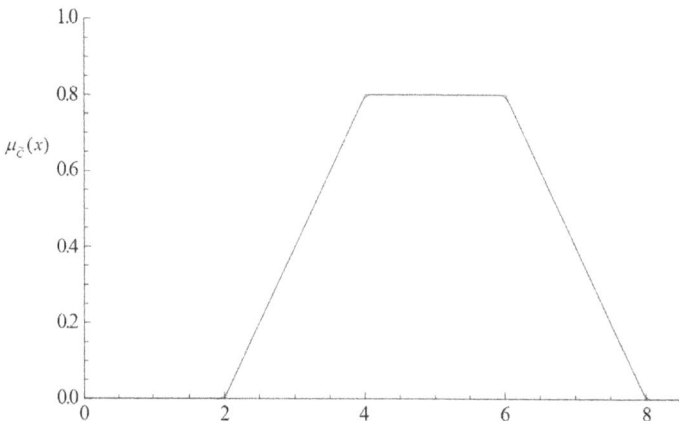

Figure 5.4 Membership function of generalized TrFN $\tilde{C}$ and its approximation.

$$\mu_{\tilde{D}}(x) = \begin{cases} 0, & if \ x < a_1; \\ \alpha_0 \dfrac{x-a_1}{a_2-a_1}, & if \ a_1 \leq \ x < a_2; \\ \alpha_0 + (1-\alpha_0)\dfrac{x-a_2}{a_3-a_2}, & if \ a_2 \leq \ x < a_3; \\ 1, & if \ a_3 \leq \ x \leq \ a_4; \\ \alpha_0 + (1-\alpha_0)\dfrac{a_4-x}{a_5-a_4}, & if \ a_4 < x \leq \ a_5; \\ \alpha_0 \dfrac{a_6-x}{a_6-a_5}, & if \ a_5 < x \leq \ a_6; \\ 0, & if \ x > a_6, \end{cases}$$

where $a_1 \leq a_2 \leq a_3 \leq a_4 \leq a_5 \leq a_6$ are real numbers and $\alpha_0 \in (0,1)$.

The membership function $\mu_{\tilde{D}}(x)$ can be represented by generalized cut functions as follows:

$$\mu_{\tilde{D}}(x) = [x]^{\alpha_0}_{a_1,a_2} + [x]^{1-\alpha_0}_{a_2,a_3} - [x]^{1-\alpha_0}_{a_4,a_5} - [x]^{\alpha_0}_{a_5,a_6},$$

so that the approximation of membership function of $\tilde{D}$ is given by

$$\mu_{\tilde{D}}^{app}(x) = S^{(\beta)}_{a_1,a_2,\alpha_0}(x) + S^{(\beta)}_{a_2,a_3,1-\alpha_0}(x) - S^{(\beta)}_{a_4,a_5,1-\alpha_0}(x) - S^{(\beta)}_{a_5,a_6,\alpha_0}(x).$$

For example, consider an α_0 -piecewise linear 1-knot fuzzy number $\tilde{D} = \tilde{D}(0.6,(1,2,4,5,7,8))$. Now with $\beta = 38$, the membership function of $\tilde{D}$ and its approximation are shown in Figure 5.5.

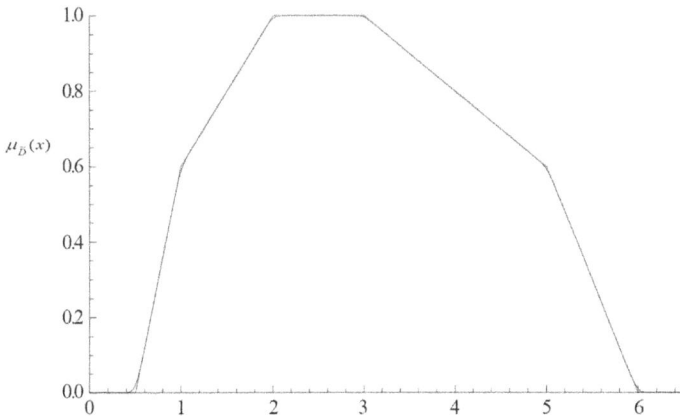

Figure 5.5 Membership function of $\tilde{D}$ and its approximation.

Note: We can use the process of approximation of piecewise linear membership functions of generalized fuzzy numbers to approximate the upper and lower membership functions of some interval type-2 fuzzy sets. For example, consider a trapezoidal interval type-2 fuzzy set (IT2 FS) $\tilde{A}$, which is represented by $\tilde{A} = (\tilde{A}^U, \tilde{A}^L) = ((a_1^U, a_2^U, a_3^U, a_4^U; w^U), (a_1^L, a_2^L, a_3^L, a_4^L; w^L))$, where $\tilde{A}^U$ denotes the upper membership function (UMF) and $\tilde{A}^L$ denotes the lower membership function (LMF) of the IT2 FS $\tilde{A}$. Here, both $\tilde{A}^U$ and $\tilde{A}^L$ are generalized trapezoidal fuzzy numbers, so their piecewise linear membership functions can be approximated to continuously differentiable membership functions.

5.5 STP WITH UNIT TRANSPORTATION COSTS REPRESENTED AS FUZZY NUMBERS

$$\text{Min } Z = \sum_{i=1}^{m} \sum_{j=1}^{n} \sum_{k=1}^{K} \tilde{c}_{ijk} x_{ijk} \tag{5.9}$$

$$\text{s.t. } \sum_{j=1}^{n} \sum_{k=1}^{K} x_{ijk} \le a_i, i = 1, 2, \ldots, m \tag{5.10}$$

$$\sum_{i=1}^{m} \sum_{k=1}^{K} x_{ijk} \ge b_j, j = 1, 2, \ldots, n \tag{5.11}$$

$$\sum_{i=1}^{m} \sum_{j=1}^{n} x_{ijk} \le e_k, k = 1, 2, \ldots, K \tag{5.12}$$

$$\sum_{i=1}^{m} a_i \ge \sum_{j=1}^{n} b_j, \sum_{k=1}^{k} e_k \ge \sum_{j=1}^{n} b_j, x_{ijk} \ge 0, \forall i, j, k$$

where $\tilde{c}_{ijk}$ are fuzzy unit transportation costs from i-th source to j-th destination via k-th conveyance, a_i denotes the amount of the product available at the i-th origin, b_j be the demand of the product of j-th destination, and e_k be transportation capacity of conveyance k.

To solve the optimization problem like defined in (5.9)–(5.12), the main difficulty is that we cannot minimize/maximize directly a fuzzy quantity. Also in case of the objective function having the coefficients as fuzzy variables with piecewise linear membership functions, one cannot apply gradient-based optimization methods directly, because a piecewise linear membership function do not have continuous gradient. This obstacle can be removed by using the continuous approximation of piecewise linear membership function as discussed in Section 5.4.

5.5.1 Solution methodology

Consider that $\tilde{c}_{ijk}$ are trapezoidal fuzzy numbers defined by $\tilde{c}_{ijk} = (c_{ijk}^1, c_{ijk}^2, c_{ijk}^3, c_{ijk}^4)$. Since $x_{ijk} \ge 0$ for all i, j, k, so $Z = \sum_{i=1}^{m} \sum_{j=1}^{n} \sum_{k=1}^{K} \tilde{c}_{ijk} x_{ijk}$

also becomes trapezoidal for any feasible solution and given by $Z = (Z_1, Z_2, Z_3, Z_4)$, where $Z_l = \sum_{i=1}^{m} \sum_{j=1}^{n} \sum_{k=1}^{K} c_{ijk}^l x_{ijk}$, $l = 1,2,3,4$.

Now we approximate trapezoidal membership function of the objective Z so that the approximation is continuously differentiable. Using (3) (cf. Section 5.4), we have the approximation of membership of Z as

$$Z' = 1/\beta \ln(\frac{1 + e^{\beta y}}{1 + e^{\beta(y-1)}}),$$ (5.13)

where

$$y = S_{Z_1,Z_2}^{(\beta)}(x) - S_{Z_3,Z_4}^{(\beta)}(x)$$

$$= \frac{1}{Z_2 - Z_1}\frac{1}{\beta}\ln(\frac{1 + e^{\beta(x-Z_1)}}{1 + e^{\beta(x-Z_2)}}) - \frac{1}{Z_4 - Z_3}\frac{1}{\beta}\ln(\frac{1 + e^{\beta(x-Z_3)}}{1 + e^{\beta(x-Z_4)}})$$ (5.14)

So one can use (13) in optimization techniques like gradient-based optimization methods. Now we construct the problems (9)–(12) using (13) by implicating that in case of $S_{Z_1,Z_2}^{(\beta)}(x)$ (which is actually approximation of $[x]_{Z_1,Z_2}$), $Z_1 \le x \le Z_2$, and in case of $S_{Z_3,Z_4}^{(\beta)}(x)$, $Z_3 \le x \le Z_4$ as follows.

$$\text{Min } Z' = \frac{1}{\beta}\ln(\frac{1 + e^{\beta y}}{1 + e^{\beta(y-1)}}),$$

$$y = t_1 - t_2,$$

$$t_1 = \frac{1}{Z_2 - Z_1}\frac{1}{\beta}\ln(\frac{1 + e^{\beta(x-Z_1)}}{1 + e^{\beta(x-Z_2)}}), Z_1 \le x \le Z_2,$$ (5.15)

$$t_2 = \frac{1}{Z_4 - Z_3}\frac{1}{\beta}\ln(\frac{1 + e^{\beta(x-Z_3)}}{1 + e^{\beta(x-Z_4)}}), Z_3 \le x \le Z_4,$$

such that the constraints (10)–(12),

$$\sum_{i=1}^{m} a_i \ge \sum_{j=1}^{n} b_j, \sum_{k=1}^{k} e_k \ge \sum_{j=1}^{n} b_j, x_{ijk} \ge 0, \forall i, j, k.$$

Solving (15) with a suitable large β, we get the values of Z_1, Z_2, Z_3, and Z_4, that is, we get the values of the fuzzy objective Z approximately equal to the TrFN (Z_1, Z_2, Z_3, Z_4) with a feasible solution $\{x_{ijk}\}$. The problem with $\tilde{c}_{ijk}$ as triangular fuzzy numbers can be constructed in the same way using the approximation of its membership function.

5.5.2 A methodology using the concept minimum of fuzzy number

In order to provide a comparison to the above methodology, here we present another method using the concept minimum of fuzzy number proposed by Buckly et al. [24]. Later, Kundu et al. [15] introduced this method to a multi-objective multi-item solid transportation problem with fuzzy parameters and obtained a fuzzy solution for the problem. In this method, the objective function with fuzzy parameters is transformed to a deterministic form. The method has the following steps to minimize a fuzzy quantity $\tilde{Z}$.

Step 1. First convert min $\tilde{Z}$ into a multi-objective problem

$$\text{Min } \tilde{Z} = (\text{Max } A_L(\tilde{Z}), \text{ Min } C(\tilde{Z}), \text{ Min } A_R(\tilde{Z})),$$

where $C(\tilde{Z})$ is the center of the core of the fuzzy number and $A_L(\tilde{Z})$ and $A_R(\tilde{Z})$ are the area under graph of the membership function of $\tilde{Z}$ to the left and right of $C(\tilde{Z})$. If the support of $\tilde{Z}$ is $[u_1, u_3]$ and the center of the core of $\tilde{Z}$ is at u_2, then

$$A_L\left(\tilde{Z}\right) = \int_{u_1}^{u_2} \mu_{\tilde{Z}}(x)dx \text{ and } A_R(\tilde{Z}) = \int_{u_2}^{u_3} \mu_{\tilde{Z}}(x)dx.$$

Step-2. Convert the above multi-objective problem into a single-objective problem as follows:

$$\text{Min } \tilde{Z} = \text{Min}\left\{\lambda_1\left[M - A_L(\tilde{Z})\right] + \lambda_2 C(\tilde{Z}) + \lambda_3 A_R(\tilde{Z})\right\}, \tag{5.16}$$

where $\lambda_l > 0$, for $l = 1, 2, 3$ and M is a large positive number so that Max $A_L(\tilde{Z})$ is equivalent to Min $\left[M - A_L(\tilde{Z})\right]$, where λ_l indicates the relative importance of $A_L(\tilde{Z}\ C(\tilde{Z})$ and $A_R(\tilde{Z})$.

Kundu et al. [15] showed that for a trapezoidal fuzzy number $\tilde{Z} = (z_1, z_2, z_3, z_4)$,

$$A_L(\tilde{Z}) = \frac{1}{2}(z_3 - z_1), C(\tilde{Z}) = \frac{z_2 + z_3}{2} \text{ and } A_R(\tilde{Z}) = \frac{1}{2}(z_4 - z_2).$$

Now applying this method to the objective function (9) of the problem (9)–(12) with $\tilde{c}_{ijk} = (c_{ijk}^1, c_{ijk}^2, c_{ijk}^3, c_{ijk}^4)$, the problem becomes

$$\text{Min } \bar{Z} = \lambda_1\left[M - A_L(Z)\right] + \lambda_2 C(Z) + \lambda_3 A_R(Z)$$

$$A_L(Z) = \frac{Z_3 - Z_1}{2}, C(Z) = \frac{Z_2 + Z_3}{2}, A_R(Z) = \frac{Z_4 - Z_2}{2} \tag{5.17}$$

subject to the constraints (10)–(12).

$$\sum_{i=1}^{m} a_i \geq \sum_{j=1}^{n} b_j, \sum_{k=1}^{k} e_k \geq \sum_{j=1}^{n} b_j$$

$$x_{ijk} \geq 0, \forall i, j, k, \lambda_1 + \lambda_2 + \lambda_3 = 1, \lambda_1 > 0, l = 1, 2, 3.$$

In order to choose the values of λ_1, it should be kept in mind that, as the above problem is a minimization problem, our aim should be more in maximizing $A_L(Z)$ (i.e., possibility of getting less value than $C(Z)$) and minimizing $C(Z)$ rather than in minimizing $A_R(Z)$ (i.e., possibility of getting more values than $C(Z)$). Also one can do a sensitivity analysis taking different set of values of λ_1 to choose suitable values of λ_1. Solving (15), we get the values of Z_1, Z_2, Z_3, and Z_4, that is, the fuzzy objective value (Z_1, Z_2, Z_3, Z_4).

5.6 NUMERICAL EXAMPLE

Now we illustrate the problem (6)–(9) with some specific data. Consider three sources, three destinations and two conveyances, that is, $i, j = 3$ and $k = 2$. The unit transportation costs $\tilde{c}_{ijk}$ are presented in Table 5.1.

$$a_1 = 38, a_2 = 30, a_3 = 40, b_1 = 32, b_2 = 36, b_3 = 35, e_1 = 50, e_2 = 56.$$

5.6.1 Solution using proposed methodology as in Section 5.1

Here we solve the problem using proposed solution methodology as described in Section 5.1. We construct the problem as (15), and with $\beta = 32$, solving this using LINGO solver, we get the following solution.

$x_{111} = 16.75, x_{131} = 21, x_{331} = 14, x_{212} = 15.25, x_{222} = 14.5, x_{322} = 21.5,$ and minimum value of
$Z = (Z_1, Z_2, Z_3, Z_4) = (345.00, 558.00, 721.00, 931.25).$

Table 5.1 Unit Transportation Costs $\tilde{c}_{ijk}$

	1	2	3	1	2	3
1	(3, 5, 7, 8)	(4, 6, 7, 10)	(3, 5, 6, 8)	(6, 8, 9, 10)	(5, 7, 8, 9)	(5, 6, 7, 9)
2	(6, 8, 9, 10)	(7, 9, 10, 11)	(6, 7, 9, 11)	(3, 5, 7, 9)	(4, 5, 6, 7)	(6, 9, 11, 13)
3	(4, 6, 8, 10)	(5, 8, 9, 11)	(3, 5, 8, 10)	(4, 7, 9, 11)	(7, 9, 10, 11)	(6, 8, 10, 12)
k		1			2	

For the fuzzy solution of the fuzzy objective function, the decision-maker gets much information about the optimum objective value. For instance, it is observed from the above result that the core of the objective value is $[558.00, 721.00]$, that is, the most possible objective value ranges between 558.00 and 721.00. Also the center of the core $((Z_2 + Z_3)/2)$ is 639.50 which may be taken as most possible crisp value of the objective function.

5.6.2 Solution using methodology as in Section 5.2

Now we solve the above problem using the concept minimum of fuzzy number as described in Section 5.2. We convert the problem as (15) using $\lambda_1 = \lambda_2 = 0.4$, $\lambda_3 = 0.2$ (as we concentrate more in maximizing $A_L(Z)$ and minimizing $C(Z)$ than in minimizing $A_R(Z)$), and $M = 400$ (a suitable large quantity so that maximizing $A_L(Z)$ is equivalent to minimizing $(M - A_L(Z))$. Using LINGO solver, based upon generalized reduced gradient (GRG) technique, we obtain the following solution. $x_{111} = 17$, $x_{131} = 21$, $x_{331} = 14$, $x_{212} = 15$, $x_{222} = 15$, and $x_{322} = 21$ so that $Z = (Z_1, Z_2, Z_3, Z_4) = (345.00, 557.00, 720.00, 930.00)$.

Remark: It is observed from above that the solutions obtained by the last two methods, that is, obtained by using the models (5.15) and (5.17), are quite similar, but it does not ensure that two methods always give similar result for any problem.

5.7 CONCLUSION

Solid transportation problem with cost coefficients of the objective function as fuzzy numbers having piecewise linear membership function (e.g., trapezoidal, triangular) is considered. Since piecewise linear membership function is not continuously differentiable, so this type of optimization problem cannot be solved directly by the gradient-based optimization techniques. In this work, this obstacle for solving such type of problems is removed by using a continuously differentiable approximation of the piecewise linear membership function. The approximation is made using interval squashing function which approximate the cut function with almost negligible error for suitable large parameter (β). With a numerical example, we successfully applied the presented method and find the solution. At the end, we may conclude that solving optimization problems with coefficients of the objective functions as fuzzy numbers like trapezoidal or triangular now becomes easy and efficient.

REFERENCES

[1] E.D. Schell, *Distribution of a Product by Several Properties*, Proceedings of 2nd Symposium in Linear Programming, DCS/comptroller, HQ US Air Force, (pp. 615–642), 1955.

[2] K. B. Haley, The sold transportation problem, *Operations Research* 10 (1962) 448–463.

[3] D. Dubois, H. Prade, *Possibility Theory: An Approach to Computerized Processing of Uncertainty*, Plenum, 1998.

[4] A. Kaufmann, *Introduction to the Theory of Fuzzy Subsets*, vol. I, Academic Press, 1975.

[5] B. Liu, *Theory and Practice of Uncertain Programming*, third ed., UTLAB, 2009. http://orsc.edu.cn/liu/up.pdf.

[6] L.A. Zadeh, Fuzzy sets, *Information Control* 8 (1965) 338–353.

[7] B. Liu, K. Iwamura, Chance constrained programming with fuzzy parameters, *Fuzzy Sets and Systems* 94(2) (1998) 227–237.

[8] P. Kundu, S. Kar, M. Maiti, Multi-objective solid transportation problems with budget constraint in uncertain environment, *International Journal of Systems Science* 45(8) (2014) 1668–1682.

[9] B. Liu, Y.K. Liu, Expected value of fuzzy variable and fuzzy expected value models, *IEEE Transactions on Fuzzy Systems* 10 (2002) 445–450.

[10] A. Das, U.K. Bera, M. Maiti, Defuzzification and application of trapezoidal type-2 fuzzy variables to green solid transportation problem, *Soft Computing* 22(7) (2018) 2275–2297.

[11] M.R. Fegad, V.A. Jadhav, A.A. Muley, Finding an optimal solution of transportation problem using interval and triangular membership functions, *European Journal of Scientific Research* 60(3) (2011) 415–421.

[12] K. Ida, M. Gen, Y. Li, *Solving Multicriteria Solid Transportation Problem with Fuzzy Numbers by Genetic Algorithms*, European Congress on Intelligent Techniques and Soft Computing (EUFIT'95), (pp. 434–441), 1995.

[13] F. Jiménez, J.L. Verdegay, Uncertain solid transportation problems, *Fuzzy Sets and Systems* 100 (1998) 45–57.

[14] F. Jiménez, J.L. Verdegay, Solving fuzzy solid transportation problems by an evolutionary algorithm based parametric approach, *European Journal of Operational Research* 117 (1999) 485–510.

[15] P. Kundu, S. Kar, M. Maiti, Multi-objective multi-item solid transportation problem in fuzzy environment, *Applied Mathematical Modelling* 37 (2013) 2028–2038.

[16] Y. Li, K. Ida, M. Gen, Improved genetic algorithm for solving multi-objective solid transportation problem with fuzzy numbers, *Computer and Industrial Engineering* 33(3–4) (1997) 589–592.

[17] N. Mathur, P.K. Srivastava, A. Paul, Algorithms for solving fuzzy transportation problem, *International Journal of Mathematics in Operational Research* 12(2) (2018) 190–219.

[18] R.J. Mitlif, M. Rasheed, S. Shihab, An optimal algorithm for a fuzzy transportation problem, *Journal of Southwest Jiaotong University* 55(3) (2020).

[19] A. Ojha, B. Das, S. Mondal, M. Maiti, An entropy based solid transportation problem for general fuzzy costs and time with fuzzy equality, *Mathematical and Computer Modeling*, 50(1–2) (2009) 166–178.

[20] S. Sadeghi-Moghaddam, M. Hajiaghaei-Keshteli, M. Mahmoodjanloo, New approaches in metaheuristics to solve the fixed charge transportation problem in a fuzzy environment, *Neural Computing and Applications* 31(1) (2019) 477–497.

[21] J. Dombi, Z. Gera, The approximation of piecewise linear membership functions and Lukasiewicz operators, *Fuzzy Sets and Systems* 154 (2005) 275–286.

[22] L. Copeland, T. Gedeon, S. Mendis, Predicting reading comprehension scores from eye movements using artificial neural networks and fuzzy output error, *Artificial Intelligence Research* 3(3) (2014) 35–48.

[23] F.G. Guimaraes, F. Campelo, R.R. Saldanha, J.A. Ramirez, A hybrid methodology for fuzzy optimization of electromagnetic devices, *IEEE Transactions on Magnetics* 41 (2005) 1744–1747.

[24] J.J. Buckley, T. Feuring, Y. Hayashi, Solving fuzzy problems in operations research: Inventory control, *Soft Computing* 7 (2002) 121–129.

[25] P. Grzegorzewski, Nearest interval approximation of a fuzzy number, *Fuzzy Sets and Systems* 130 (2002) 321–330.

[26] R. Cignoli, I.M.L. D'Ottaviano, D. Mundici, Algebraic foundations of many-valued reasoning, *Trends in Logic* 7 (2000).

[27] P. Hájek, *Metamathematics of Fuzzy Logic*, Kluwer, 1998.

[28] L. Coroianu, M. Gagolewski, P. Grzegorzewski, M.A. Firozja, T. Houlari, Piecewise linear approximation of fuzzy numbers preserving the support and core, in: Laurent, A. et al. (Eds.), *Information Processing and Management of Uncertainty in Knowledge-Based Systems, Part II, (CCIS 443)*, (pp. 244–254), 2014, Springer.

Chapter 6

Profit maximization inventory model for non-instantaneous deteriorating items with imprecise costs

D. K. Nayak, S. K. Paikray, and A. K. Sahoo

6.1 INTRODUCTION

The business organizations and industries perform many daily activities for smooth running of their business affairs in achieving the goals. The inventory management is the one of the such activities. It takes care of the amount of replenishment items and the ordering time. While developing the inventory models, researchers consider many constraints such as demand, deterioration, warehouse facility, ordering cost, holding cost, inflation, shortages, backlogging, trade credit, and advertisement cost.

In this physical world, almost all the products deteriorate over time. For instance, the agricultural products, food items, vegetables, and fruits have short life span, after that they deteriorate over the time. Gasoline, alcohol and many liquid substances have a low deterioration rate in the form of evaporation. Similarly, the electronic goods, radioactive substances, and photographic film are perishable after their life span. But the fashionable products like garments, jewelry, and cosmetics will obsolete over the time or on the introduction of new items. Thus, the deteriorating nature of the products doesn't allow to store them for longer time. On the other hand, the demand pattern of various items is different and depends on several factors like consumer, business cycle, and market place. Thus, many researchers discussed different types of demand patterns in their works, such as, demand depends on time, advertisement cost, selling price, and trade credit financing. All these demands may increase, decrease or constant throughout the cycle. Thus, the stock level of inventory depends on the demand. That is, the demand and deterioration of items play a major role in inventory management.

The total inventory cost includes holding cost, ordering cost, deterioration cost, interest cost of investment, etc. Mostly, all these costs are basically considered as deterministic while developing the mathematical model. However, in general these costs are fluctuating due to several socio-economic factors. For example, the ordering cost may fluctuate due to fluctuation

DOI: 10.1201/9781003462422-6

in petrochemicals price. Also, the change in tax price and amendments of government or company policies affects different costs of inventory. That means, most of the inventory costs are of the imprecise nature in the real market environments.

In recent years, a numerous research articles were published on inventory problems for items with different demand conditions under several inventory constraints. The inventory items having price dependent demand was discussed by Barik et al. [1]. The time-dependent demand inventory problems were considered by Barik et al. [2, 3], Indrajitsingha et al. [4], Routray et al. [5], and Kumar et al. [6]. Similarly, Shaikh et al. [7] developed an inventory model for stock-dependent demand items. But Chen et al. [8] formulated an inventory model for items with stock-level-dependent, time-varying, and price-dependent demand. Chakraborty et al. [9] found an optimal strategy for inventory problem for ramp-type demand items. For more inventory problems with different demands, one can refer the current works of Wu et al. [10] for trapezoidal-type demand, Jaggi et al. [11] for exponential declining demand, Kumar et al. [12] for fuzzy demand, and Shaikh et al. [13] for demand depends on both price and frequency of advertisement.

Most of the researchers represent the inventory constraints with deterministic parameters based on the existing past data. But in reality, this data is inadequate to represent the present constraints due to different changes in the competitive business and technological advancement. That is, the values of parameters involved in inventory model are fluctuating within a limit due to several reasons. As a result, the use of deterministic parameters and classical approach for solution of inventory problems of present scenario is no longer useful. Thus to overcome these problems, many researchers applied the fuzzy set theory and their applications to obtain the optimal strategy for different inventory problems. In the last decade, Sharmila and Uthayakumar [14], Kumar and Rajput [15], Sen et al. [16], Sangal et al. [17], Saha [18], Saha and Chakrabarti [19], and Routray et al. [5] dealt their inventory problems in fuzzy environment. In this context, we refer the interested readers to the current works of Shaikh et al. [20], Kumar et al. [6, 12, 21, 22, 23], Nayak et al. [24], and Padhy et al. [25].

Motivated essentially by the aforementioned works, here we develop an inventory model for deteriorating items which follows the exponential demand pattern in the first interval and constant demand in the second interval. The associated costs are assumed to be imprecise in nature with relevant to the actual market scenario. The main objective of this study is to maximize the total profit of the inventory problem. Thus, we present a mathematical approach for the solution of the model in fuzzy environment. In particular, the triangular fuzzy numbers are used to express the imprecise costs and graded mean integration representation method is employed for defuzzification of the resultant fuzzy functions. Furthermore, the numerical exercise of the proposed model is illustrated via different examples. Finally, the sensitivity behavior of parameters is examined to draw the managerial decisions.

6.2 ASSUMPTIONS AND NOTATIONS

The proposed inventory model has been developed on the basis of following assumptions and notations.

6.2.1 Assumptions

The following assumptions are taken into consideration in formulating the model:

1. The inventory system is considered for single items only and the planning horizon is infinite.
2. Replenishment rate occurs instantaneously at an infinite rate.
3. The demand rate at any time t is

$$D(t) = \begin{cases} \lambda e^{\mu t}, & 0 < t < m, \\ \lambda, & m < t < T. \end{cases} \text{ Here, } \lambda > 0 \text{ and } \mu \neq 0.$$

4. The time variable deterioration rate is $\dfrac{1}{1+(L-t)}$, $m < t < T$.
5. The associated inventory costs are considered as imprecise and are represented by fuzzy parameters.
6. The life span of the items is greater than that of total cycle time.

6.2.2 Notations

The different notations considered for various parameters of inventory are as follows:

1. $V(t)$: The stock level of items in the inventory at any time t.
2. S: The optimal ordering quantity (i.e., $S = V(0)$).
3. T: The time length of the inventory cycle.
4. L: The life span of the item.
5. i: The inventory carrying charge as fraction per unit time.
6. $\tilde{A}$: Fuzzy ordering cost.
7. $\tilde{p}$: Fuzzy purchase cost per unit item.
8. $\tilde{S}$: Fuzzy selling price per unit item.
9. PC: Total purchase cost of items in a cycle.
10. HC: Total holding cost of inventory per cycle.
11. OC: Ordering cost per cycle.
12. DU: Total number of deteriorated items in the cycle.

13. DC: The total deterioration cost of items in a business cycle.
14. TC: The total average cost of inventory per unit time.
15. SR: The average sales revenue per unit time.
16. $\Pr(m,T)$: The average profit per unit time.
17. $\mathrm{PR}_G(m,T)$: The average defuzzified profit per unit time.
18. m^*: Optimal value of m.
19. T^*: Optimal value of T.
20. $\mathrm{PR}_G^*(m,T)$: Optimal average profit per unit time.

6.3 MATHEMATICAL FORMULATION

The business cycle starts at $t = 0$ with S number of inventory items and ends at time T, where the inventory is empty. During the business cycle, at any time t such that $0 < t < m$, the inventory level depletes due to exponential demand and at any time t such that $m < t < T$, the inventory level depletes due to both constant demand and time variable deterioration. The mathematical form of inventory level at any time in the business cycle can be expressed by the following differential equations.

$$\frac{dV(t)}{dt} = -\lambda e^{\mu t} \tag{6.1}$$

with the condition $S = V(0)$, and

$$\frac{dV(t)}{dt} + \frac{1}{1+(L-t)}V(t) = -\lambda \tag{6.2}$$

with the condition $V(T) = 0$.
Solution of equation (1) with the initial condition is

$$V(t) = S + \frac{\lambda}{\mu}\left(1 - e^{\mu t}\right), 0 \le t \le m. \tag{6.3}$$

Similarly, the solution of equation (2) with the boundary condition is

$$V(t) = \lambda(1 + L - t)\log\left[\frac{1+(L-t)}{1+(L-T)}\right]. \tag{6.4}$$

The inventory level at time $t = m$ can be calculated from equation (3) as

$$V(m) = S + \frac{\lambda}{\mu}\left(1 - e^{\mu m}\right), 0 \le t \le m. \tag{6.5}$$

The inventory level at time $t = m$ can be calculated from equation (4) as

$$V(m) = \lambda(1 + L - m)\log\left[\frac{1+(L-m)}{1+(L-T)}\right]. \tag{6.6}$$

Solving equations (5) and (6) for S, we obtained

$$S = \lambda(1 + L - m)\log[\frac{1 + (L - m)}{1 + (L - T)}] - \frac{\lambda}{\mu}(1 - e^{\mu m}). \tag{6.7}$$

Total number of deteriorated units in the complete cycle is

$$DU = V(0) - \left[\int_0^m D(t) \, dt + \int_m^T D(t) \, dt\right]$$

$$= \lambda(1 + L - m)\log[\frac{1 + (L - m)}{1 + (L - T)}] - \lambda T + \lambda m. \tag{6.8}$$

6.3.1 Different inventory costs

Since the cost parameters are assumed to be fuzzy, the associated costs are also fuzzy and are calculated as follows:

Purchase Cost

The total purchase cost of items in inventory cycle is

$$PC = \tilde{p} * S$$

$$= \tilde{p} * [\lambda(1 + L - m)\log[\frac{1 + (L - m)}{1 + (L - T)}] - \frac{\lambda}{\mu}(1 - e^{\mu m})]. \tag{6.9}$$

Holding Cost

The total holding cost of inventory per cycle is

$$HC = \tilde{p} * i\left[\int_0^m V(t) \, dt + \int_m^T V(t) \, dt\right]$$

$$= \tilde{p} * i[\int_0^m \{S + \frac{\lambda}{\mu}(1 - e^{\mu t})\} \, dt + \int_m^T \{\lambda(1 + L - t)\log[\frac{1 + (L - t)}{1 + (L - T)}]\} \, dt]$$

$$= \tilde{p} * i[\lambda\{((1 + L - m)\log[\frac{1 + (L - m)}{1 + (L - T)}] - \frac{1}{\mu}(1 - e^{\mu m}))m$$

$$+ \frac{1}{\mu}\left(m - \frac{e^{\mu m}}{\mu}\right) + \frac{1}{\mu^2}\}$$

$$+ \frac{\lambda}{2}[(1 + L - m)^2\log(1 + L - m) - (1 + L - T)^2\log(1 + L - T)$$

$$+ \frac{1}{2}(1 + L - T)^2$$

$$- \frac{1}{2}(1 + L - m)^2] - \{\lambda\log(1 + L - T)[(T + LT - \frac{T^2}{2})$$

$$- (m + Lm - \frac{m^2}{2})]\}]. \tag{6.10}$$

Deterioration Cost

The total deterioration cost of inventory per cycle per is

$$DU = \tilde{p} * DU = \tilde{p} * \left[\lambda (1 + L - m) \log \left[\frac{1 + (L - m)}{1 + (L - T)} \right] - \lambda T + \lambda m \right]. \quad (6.11)$$

Ordering Cost

Ordering cost per cycle is

$$OC = \tilde{A}. \quad (6.12)$$

Total Cost

The total cost of inventory per cycle is the sum of ordering cost, purchase cost, holding cost, and deteriorating cost, whereas the total average cost of inventory is calculated as

$$TC = \frac{1}{T} * [OC + PC + HC + DC]$$

$$= \frac{1}{T} * [\tilde{A} + \tilde{p} * [\lambda(1 + L - m) \log[\frac{1 + (L - m)}{1 + (L - T)}] - \frac{\lambda}{\mu}(1 - e^{\mu m})]$$

$$+ \tilde{p} * i[\lambda\{((1 + L - m) \log[\frac{1 + (L - m)}{1 + (L - T)}] - \frac{1}{\mu}(1 - e^{\mu m}))m$$

$$+ \frac{1}{\mu}(m - \frac{e^{\mu m}}{\mu}) + \frac{1}{\mu^2}\} + \frac{\lambda}{2}[(1 + L - m)^2 \log(1 + L - m)$$

$$- (1 + L - T)^2 \log(1 + L - T) + \frac{1}{2}(1 + L - T)^2 - \frac{1}{2}(1 + L - m)^2]$$

$$- \{\lambda\log(1 + L - T)[(T + LT - \frac{T^2}{2}) - (m + Lm - \frac{m^2}{2})]\}$$

$$+ \tilde{p} * [\lambda(1 + L - m) \log[\frac{1 + (L - m)}{1 + (L - T)}] - \lambda T + \lambda m]]. \quad (6.13)$$

6.3.2 Sales revenue

The total average sales revenue is calculated as

$$SR = \frac{1}{T} * [\tilde{s} * \text{Number of sales unit}]$$

$$= \frac{1}{T} * [\tilde{s} * (\text{Initial Inventory} - \text{Deteriorated Units})]$$

$$= \frac{1}{T} * [\tilde{s} * (s - DU)]$$

$$= \frac{1}{T} * [\tilde{s} * \{(\lambda(1+L-m)\log[\frac{1+L-m}{1+L-T}] - \frac{\lambda}{\mu}(1-e^{\mu m}))$$

$$-(\lambda(1+L-m)\log[\frac{1+L-m}{1+L-T}]$$

$$-\lambda T + \lambda m)\}]$$

$$= \frac{1}{T} * [\tilde{s} * \{-\frac{\lambda}{\mu} + \frac{\lambda}{\mu}e^{\mu m} + \lambda T - \lambda m\}]. \tag{6.14}$$

6.3.3 Profit function

As the inventory costs calculated are fuzzy, the total average profit function is also fuzzy and it can be calculated as

$$Pr(m,T) = SR - TC$$

$$= \frac{1}{T} * \{[\tilde{s} * \{-\frac{\lambda}{\mu} + \frac{\lambda}{\mu}e^{\mu m} + \lambda T - \lambda m\}] - [\tilde{A}$$

$$+ \tilde{p} * [\lambda(1+L-m)\log[\frac{1+(L-m)}{1+(L-T)}]$$

$$-\frac{\lambda}{\mu}(1-e^{\mu m})] + \tilde{p} * i[\lambda\{((1+L-m)\log\left[\frac{1+(L-m)}{1+(L-T)}\right]$$

$$-\frac{1}{\mu}(1-e^{\mu m}))m$$

$$+\frac{1}{\mu}\left(m-\frac{e^{\mu m}}{\mu}\right)+\frac{1}{\mu^2}\} + \frac{\lambda}{2}[(1+L-m)^2\log(1+L-m)$$
$$-(1+L-T)^2\log(1+L-T)$$

$$+\frac{1}{2}(1+L-T)^2 - \frac{1}{2}(1+L-m)^2]$$

$$-\{\lambda\log(1+L-T)[(T+LT-\frac{T^2}{2})$$

$$-\left(m+Lm-\frac{m^2}{2}\right)]\}$$

$$+ \tilde{p} * [\lambda(1+L-m)\log[\frac{1+(L-m)}{1+(L-T)}] - \lambda T + \lambda m]]\}. \tag{6.15}$$

6.3.4 Defuzzification

Let the fuzzy parameters considered for different costs are characterized by triangular fuzzy numbers. That is, $\tilde{A} = (A_1, A_2, A_3)$, $\tilde{p} = (p_1, p_2, p_3)$, and $\tilde{s} = (s_1, s_2, s_3)$.

Now, the fuzzy profit function in equation (15) is defuzzified as follow:
Replacing $\tilde{A}$ by A_1, $\tilde{p}$ by p_1, and $\tilde{s}$ by s_1, we get

$$
Pr_1(m,T) = \frac{1}{T} * \{[s_1 * \{-\frac{\lambda}{\mu} + \frac{\lambda}{\mu}e^{\mu m} + \lambda T - \lambda m\}]
$$

$$
-[A_1 + p_1 * [\lambda(1 + L - m)\log[\frac{1+(L-m)}{1+(L-T)}]
$$

$$
-\frac{\lambda}{\mu}(1 - e^{\mu m})] + p_1 * i[\lambda\{(1 + L - m)\log\left[\frac{1+(L-m)}{1+(L-T)}\right] - \frac{1}{\mu}(1 - e^{\mu m})\right)m
$$

$$
+ \frac{1}{\mu}\left(m - \frac{e^{\mu m}}{\mu}\right) + \frac{1}{\mu^2}\} + \frac{\lambda}{2}[(1 + L - m)^2\log(1 + L - m)
$$

$$
-(1 + L - T)^2\log(1 + L - T)
$$

$$
+ \frac{1}{2}(1 + L - T)^2 - \frac{1}{2}(1 + L - m)^2] - \{\lambda\log(1 + L - T)[(T + LT - \frac{T^2}{2})
$$

$$
-\left(m + Lm - \frac{m^2}{2}\right)]\}
$$

$$
+ p_1 * [\lambda(1 + L - m)\log[\frac{1+(L-m)}{1+(L-T)}] - \lambda T + \lambda m]]\}. \qquad (6.16)
$$

Replacing $\tilde{A}$ by A_2, $\tilde{p}$ by p_2, and $\tilde{s}$ by s_2, we get

$$
Pr_2(m,T) = \frac{1}{T} * \left\{s_2 * \left\{-\frac{\lambda}{\mu} + \frac{\lambda}{\mu}e^{\mu m} + \lambda T - \lambda m\right\}\right] - [A_2
$$

$$
+ p_2 * \left[\lambda(1 + L - m)\log\left[\frac{1+(L-m)}{1+(L-T)}\right]\right]
$$

$$
-\frac{\lambda}{\mu}(1 - e^{\mu m})] + p_2 * i[\lambda\{(1 + L - m)\log\left[\frac{1+(L-m)}{1+(L-T)}\right]
$$

$$
-\frac{1}{\mu}(1 - e^{\mu m})\right)m
$$

$$
+ \frac{1}{\mu}\left(m - \frac{e^{\mu m}}{\mu}\right) + \frac{1}{\mu^2}\} + \frac{\lambda}{2}[(1 + L - m)^2\log(1 + L - m)
$$

$$
-(1 + L - T)^2\log(1 + L - T)
$$

$$+ \frac{1}{2}(1+L-T)^2 - \frac{1}{2}(1+L-m)^2]$$

$$-\{\lambda\log(1+L-T)[\left(T+LT-\frac{T^2}{2}\right)$$

$$-\left(m+Lm-\frac{m^2}{2}\right)]\}+p_2*[\lambda(1+L-m)$$

$$\log\left[\frac{1+(L-m)}{1+(L-T)}\right]-\lambda T+\lambda m]]\}. \tag{6.17}$$

Replacing $\tilde{A}$ by A_3, $\tilde{p}$ by p_3, and $\tilde{s}$ by s_3, we get

$$Pr_3(m,T)=\frac{1}{T}*\{[s_3*\{-\frac{\lambda}{\mu}+\frac{\lambda}{\mu}e^{\mu m}+\lambda T-\lambda m\}]$$

$$-[A_3+p_3*[\lambda(1+L-m)\log\frac{1+(L-m)}{1+(L-T)}]$$

$$-\frac{\lambda}{\mu}(1-e^{\mu m})]+p_3*i[\lambda\{((1+L-m)\log\left[\frac{1+(L-m)}{1+(L-T)}\right]$$

$$-\frac{1}{\mu}(1-e^{\mu m}))m$$

$$+\frac{1}{\mu}\left(m-\frac{e^{\mu m}}{\mu}\right)+\frac{1}{\mu^2}\}+\frac{\lambda}{2}[(1+L-m)^2\log(1+L-m)$$

$$-(1+L-T)^2\log(1+L-T)$$

$$+\frac{1}{2}(1+L-T)^2-\frac{1}{2}(1+L-m)^2]$$

$$-\{\lambda\log(1+L-T)[\left(T+LT-\frac{T^2}{2}\right)$$

$$-\left(m+Lm-\frac{m^2}{2}\right)]\}+p_3*[\lambda(1+L-m)$$

$$\log\left[\frac{1+(L-m)}{1+(L-T)}\right]-\lambda T+\lambda m]]\}. \tag{6.18}$$

Using equations (16)–(18) and graded mean integration representation (GMIR) method, the defuzzified total average profit function can be obtained as follows:

$$PR_G(m,T)=\frac{1}{6}[Pr_1(m,T)+4*Pr_2(m,T)+Pr_3(m,T)]$$

That is,

$$
\begin{aligned}
\mathrm{PR}_G\left(m,T\right) = \frac{1}{T} * \{ & [(s_1 + 4*s_2 + s_3) * \left\{ -\frac{\lambda}{\mu} + \frac{\lambda}{\mu} e^{\mu m} + \lambda T - \lambda m \right\}] \\
& -[(A_1 + 4*A_2 + A_3) \\
& + (p_1 + 4*p_2 + p_3) * \left[\lambda(1 + L - m) \log \left[\frac{1+(L-m)}{1+(L-T)} \right] \right. \\
& \left. -\frac{\lambda}{\mu}(1 - e^{\mu m}) \right] \\
& + (p_1 + 4*p_2 + p_3) * i[\lambda \{ (1 + L - m) \log \left[\frac{1+(L-m)}{1+(L-T)} \right] \\
& -\frac{1}{\mu}(1 - e^{\mu m}) \bigg] m \\
& + \frac{1}{\mu}\left(m - \frac{e^{\mu m}}{\mu} \right) + \frac{1}{\mu^2} \} + \frac{\lambda}{2}[(1 + L - m)^2 \log(1 + L - m) \\
& -(1 + L - T)^2 \log(1 + L - T) \\
& + \frac{1}{2}(1 + L - T)^2 - \frac{1}{2}(1 + L - m)^2] \\
& -\{ \lambda \log(1 + L - T)[(T + LT - \frac{T^2}{2}) - (m + Lm \\
& -\frac{m^2}{2})]\} + (p_1 + 4*p_2 + p_3) * [\lambda(1 + L - m) \\
& \log \left[\frac{1+(L-m)}{1+(L-T)} \right] - \lambda T + \lambda m]]\}.
\end{aligned}
\tag{6.19}
$$

6.4 SOLUTION PROCEDURE

Our objective is to find the maximum value of total average profit function in equation (19). For which, we need to proceed as follows:

Find m^* such that $\dfrac{\partial \mathrm{PR}_G\left(m,T\right)}{\partial m} = 0$.

Find T^* such that $\dfrac{\partial \mathrm{PR}_g\left(m,T\right)}{\partial T} = 0$.

Find $\dfrac{\partial^2 \mathrm{PR}_G\left(m,T\right)}{\partial m^2}$, $\dfrac{\partial^2 \mathrm{PR}_G\left(m,T\right)}{\partial T^2}$, and $\dfrac{\partial^2 \mathrm{PR}_G\left(m,T\right)}{\partial m \partial T}$ atm^* and T^* .

Check if

$$\frac{\partial^2 PR_G(m,T)}{\partial m^2} < 0, \ \frac{\partial^2 PR_G(m,T)}{\partial T^2} < 0, \ \frac{\partial^2 PR_G(m,T)}{\partial m \partial T} > 0 \text{ and}$$

$$(\frac{\partial^2 PR_G(m,T)}{\partial m^2} * \frac{\partial^2 PR_G(m,T)}{\partial T^2} - \left(\frac{\partial^2 PR_G(m,T)}{\partial m \partial T}\right)^2) > 0 \text{ at } m^* \text{ and } T^*,$$

then $\{m^*, T^*\}$ is the optimal solution.

Using m^* for m and T^* for T in equations (7) and (19), find the initial inventory S and average profit per unit time $PR_G(m,T)$.

6.5 NUMERICAL EXAMPLES

Based on the aforementioned solution procedure and by using the Mathematica 11.1.1 software, the following examples are illustrated to validate our proposed model.

Example 1: The values of different parameters involved in inventory are $\lambda = 15.2$, $\mu = 0.22$, $i = 0.01$, $L = 24.5$, $\tilde{A} = (A_1, A_2, A_3) = (90, 103, 116)$, $\tilde{p} = (p_1, p_2, p_3) = (10, 15.2, 20)$, and $\tilde{s} = (s_1, s_2, s_3) = (15.2, 20.2, 25.2)$.

Solution:
The optimal solution is:

$$m^* = 0.453374 \quad T^* = 2.03928 \quad S = 32.1495 \quad PR_G(m,T) = 13.1478.$$

Example 2: The values of different parameters involved in inventory are $\lambda = 15$, $\mu = 0.21$, $i = 0.01$, $L = 25$, $\tilde{A} = (A_1, A_2, A_3) = (105, 105.5, 106)$, $\tilde{p} = (p_1, p_2, p_3) = (15, 15.1, 15.6)$, and $\tilde{s} = (s_1, s_2, s_3) = (19.7, 20, 20.2)$.

Solution:
The optimal solution is:

$$m^* = 1.3003, \ T^* = 2.12362, \ S = 34.9877, \text{ and } PR_G(m,T) = 23.5707.$$

Example 3: The values of different parameters involved in inventory are $\lambda = 15.5$, $\mu = 0.21$, $i = 0.01$, $L = 25$, $\tilde{A} = (A_1, A_2, A_3) = (105, 105.5, 110)$, $\tilde{p} = (p_1, p_2, p_3) = (15, 15.5, 16)$, and $\tilde{s} = (s_1, s_2, s_3) = (15, 20, 25)$.

Solution:
The optimal solution is:

$$m^* = 2.05127, \ T^* = 3.63377, \ S = 65.1185, \text{ and}$$
$$PR_G(m,T) = 38.4394.$$

Example 4: The values of different parameters involved in inventory are $\lambda = 15.5$, $\mu = 0.22$, $i = 0.01$, $L = 24$, $\tilde{A} = (A_1, A_2, A_3) = (98, 107.5, 116)$, $\tilde{p} = (p_1, p_2, p_3) = (9.2, 15.2, 21.2)$, and $\tilde{s} = (s_1, s_2, s_3) = (15, 20.1, 25)$.

Solution
The optimal solution is:

$m^* = 1.03291$, $T^* = 2.41672$, $S = 38.7758$, and
$PR_G(m, T) = 21.8166$.

Example 5: The values of different parameters involved in inventory are $\lambda = 15.5$, $\mu = 0.21$, $i = 0.01$, $L = 25$, $\tilde{A} = (A_1, A_2, A_3) = (105, 110, 115)$, $\tilde{p} = (p_1, p_2, p_3) = (15, 16, 17)$, and $\tilde{s} = (s_1, s_2, s_3) = (19, 20, 21)$.

Solution
The optimal solution is:

$m^* = 3.64955$, $T^* = 3.8546$, $S = 85.3768$, and $PR_G(m, T) = 52.425$.

6.6 SENSITIVITY ANALYSIS

The sensitivity behavior of parameters λ and μ are illustrated numerically in Tables 6.1 and 6.2 and graphically in Figures 6.1 and 6.2 (*Example 3* is considered).

Table 6.1 Effect of Change in Values of Parameters of Optimal Results

λ	m	T	S	$PR_G(m, T)$
15.5	2.05127	3.63377	65.1185	35.4394
16.5	1.96494	4.01351	75.4635	39.1649
17.5	1.92751	4.03142	80.1084	42.2044
18.5	1.92086	4.06392	85.2954	45.7122
19.5	2.02661	3.51029	79.1256	55.9813

Table 6.2 Effect of Change in Values of Parameters of Optimal Results

λ	m	T	S	$PR_G(m, T)$
0.20	3.00685	2.35593	53.9588	51.0463
0.21	2.05127	3.63377	65.1185	35.4394
0.22	1.4806	3.54947	60.6297	28.264
0.23	1.09685	2.54682	42.4933	21.0028
0.24	0.690358	2.31273	37.6271	11.864

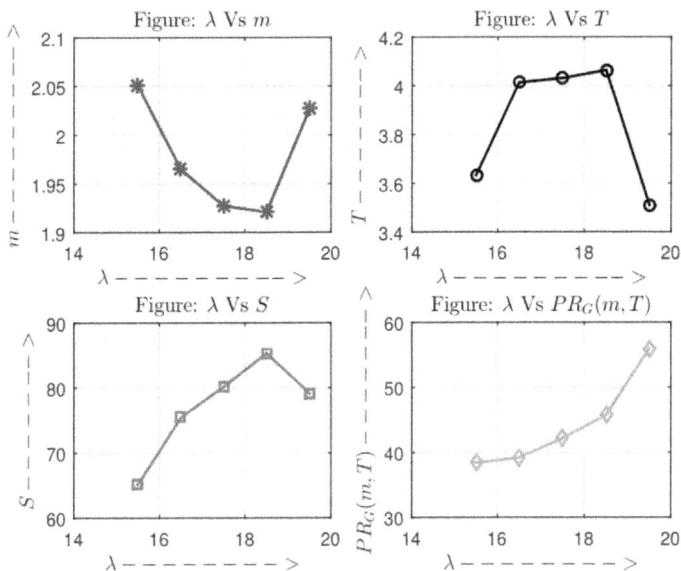

Figure 6.1 Sensitivity effect of λ on optimal results.

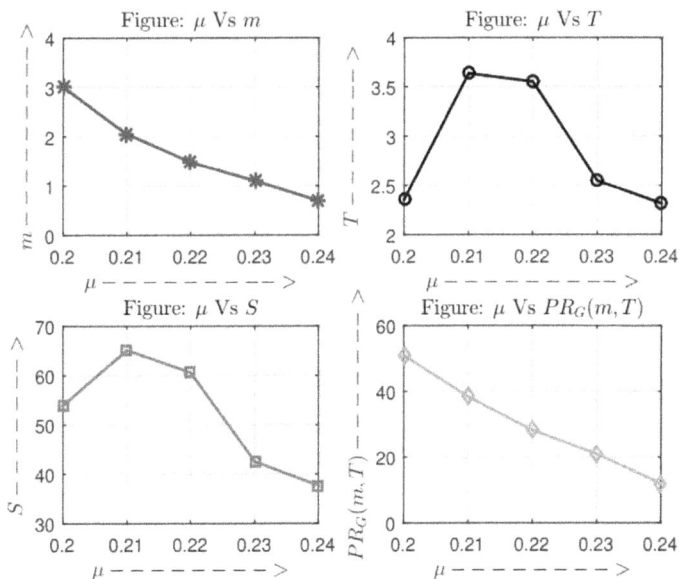

Figure 6.2 Sensitivity effect of μ on optimal results.

6.7 CONCLUSION

In the real business market, there are several non-instantaneous deteriorating items which doesn't follow the uniform demand throughout the cycle. Also, the different inventory costs have imprecise nature due to many socioeconomic factors. Thus, in this study, we considered the non-instantaneous items which have exponential demand with no deterioration in the first interval and constant demand with variable deterioration in the second interval. Mainly, the associated costs with inventory are considered imprecise nature, unlike in classical inventory models. Therefore, the inventory model is developed in fuzzy environment to obtain the optimal strategy which is more suitable to the real-world inventory problem. In fuzzy environment, the triangular fuzzy numbers are used to characterize the imprecise nature of the costs and the Graded Mean Integration Representation method is implemented to defuzzify the resultant fuzzy functions. Moreover, the mathematical approach for obtaining the optimal profit is illustrated with numerical examples. Finally, the sensitivity analysis of important parameters is carried out to help the inventory decision-makers. Furthermore, the model can be extended by incorporating the shortages, inflation, trade credit financing, and many other constraints.

REFERENCES

[1] S. Barik, S. Mishra, S. K. Paikray and U. K. Misra, A deteriorating inventory model with shortages under price dependent demand and inflation, *Asian J. Math. Comput. Res.* **2016** (2016), 14–25.

[2] S. Barik, S. K. Paikray and U. K. Misra, Inventory model of deteriorating items for nonlinear holding cost with time dependent demand, *J. Adv. Math.* **9** (2014), 2705–2709.

[3] S. Barik, S. Mishra, S. K. Paikray and U. K. Misra, An inventory model for deteriorating items under time varying demand condition, *Int. J. Appl. Eng. Res.* **10** (2015), 35770–35773.

[4] S. K. Indrajitsingha, S. S. Routray, S. K. Paikray and U. Misra, Fuzzy economic production quantity model with time dependent demand rate, *Log Forum* **12** (2016), 193–198 [https://doi.org/10.17270/j.log.2016.3.1].

[5] S. S. Routray, S. K. Paikray, S. Mishra and U. K. Misra, Fuzzy inventory model with single item under time dependent demand and holding cost, *Int. J. Adv. Res. Sci.* **6** (2017), 1604–1618.

[6] B. A. Kumar, S. K. Paikray and U. Misra, Two-Storage fuzzy inventory model with time dependent demand and holding cost under acceptable delay in payment, *Math. Model. Anal.* **25** (2020), 441–460 [https://doi.org/10.3846/mma.2020.10805].

[7] A. A. Shaikh, L. E. Cárdenas-Barrón and S. Tiwari, A two-warehouse inventory model for non-instantaneous deteriorating items with interval valued inventory costs and stock dependent demand under inflationary conditions, *Neural Comput. Appl.* **31** (2019), 1931–1948 [https://doi.org/10.1007/s00521-017-3168-4].

[8] L. Chen, X. Chen, M. F. Keblis and G. Li, Optimal pricing and replenishment policy for deteriorating inventory under stock-level-dependent, time-varying and price-dependent demand, *Comput. Ind. Eng.* **135** (2019), 1294–1299 [https://doi.org/10.1016/j.cie.2018.06.005].

[9] D. Chakraborty, D. K. Jana and T. K. Roy, Two-warehouse partial backlogging inventory model with ramp type demand rate, three-parameter Weibull distribution deterioration under inflation and permissible delay in payments, *Comput. Ind. Eng.* **123** (2018), 157–179 [https://doi.org/10.1016/j.cie.2018.06.022].

[10] J. Wu, J-T. Teng, and K. Skouri, Optimal inventory policies for deteriorating items with trapezoidal-type demand patterns and maximum lifetimes under upstream and downstream trade credits. *Ann. Oper. Res.* **264** (2018), 459–476 [https://doi.org/10.1007/s10479-017-2673-2].

[11] C. K. Jaggi, P. Gautam and A. Khanna, Inventory decisions for imperfect quality deteriorating items with exponential declining demand under trade credit and partially backlogged shortages, in: P. Kapur, U. Kumar and A. Verma (Eds.), *Quality, IT and Business Operations*, Springer Proceedings in Business and Economics, Singapore, pp. 213–229, 2018 [https://doi.org/10.1007/978-981-10-5577-5_18].

[12] B. A. Kumar, S. K. Paikray and H. Dutta, Cost optimization model for items having fuzzy demand and deterioration with two-warehouse facility under the trade credit financing, *AIMS Math.* **5** (2020), 1603–1620 [https://doi.org/10.3934/math.2020109].

[13] A. A. Shaikh, L. E. Cárdenas-Barrón, A. K. Bhunia and S. Tiwari, An inventory model of a three parameter Weibull distributed deteriorating item with variable demand dependent on price and frequency of advertisement under trade credit, *RAIRO Oper. Res.* **53** (2019), 903–916 [https://doi.org/10.1051/ro/2017052].

[14] D. Sharmila and R. Uthayakumar, Inventory model for deteriorating items involving fuzzy with shortages and exponential demand, *Int. J. Supply Oper. Manag.* **2** (2015), 888–904.

[15] S. Kumar and U. S. Rajput, Fuzzy inventory model for deteriorating items with time dependent demand and partial backlogging, *Appl. Math.*, **6** (2015), 496–509.

[16] N. Sen, B. K. Nath and S. Saha, A fuzzy inventory model for deteriorating items based on different defuzzification techniques, *Amer. J. Math. Stat.* **6** (2016), 128–137.

[17] I. Sangal, A. Agarwal, and S. Rani, A fuzzy environment inventory model with partial backlogging under learning effect, *Int. J. Comput. Appl.* **137** (2016), 25–32.

[18] S. Saha, Fuzzy inventory model for deteriorating items in a supply chain system with price dependent demand and without backorder, *Amer. J. Eng. Res.* **6** (2017), 183–187.

[19] S. Saha and T. Chakrabarti, A fuzzy inventory model for deteriorating items with linear price dependent demand in a supply chain, *Intern. J. Fuzzy Math. Arch.* **13** (2017), 59–67.

[20] A. A. Shaikh, L. E. Cárdenas-Barrón and L. Sahoo, A fuzzy inventory model for a deteriorating item with variable demand, permissible delay in payments and partial backlogging with shortage follows inventory (SFI) policy, *Int. J. Fuzzy Syst.* **20** (2018), 1606–1623 [https://doi.org/10.1007/s40815-018-0466-7].

[21] B. A. Kumar, S. K. Paikray, S. Mishra and S. Routray, A fuzzy inventory model of defective items under the effect of inflation with trade credit financing, in: O. Castillo, D. Jana, D. Giri and A. Ahmed (Eds.), *Recent Advances in Intelligent Information Systems and Applied Mathematics, ICITAM 2019: Studies in Computational Intelligence*, vol. 863, Springer, Singapore, 2019 [https://doi.org/10.1007/978-3-030-34152-7_62].

[22] B. A. Kumar, S. K. Paikray and B. Padhy, Retailer's optimal ordering policy for deteriorating inventory having positive lead time under pre-payment interim and post-payment strategy, *Int. J. Appl. Comput. Math.* 8 (2022), 1–33.

[23] B. A. Kumar and S. K. Paikray Cost optimization inventory model for deteriorating items with trapezoidal demand rate under completely backlogged shortages in crisp and fuzzy environment, *RAIRO Oper. Res.* 56 (2022), 1969–1994.

[24] D. K. Nayak, S. S. Routray, S. K. Paikray and H. Dutta, A fuzzy inventory model for Weibull deteriorating items under completely backlogged shortages, *Discrete Contin. Dyn. Syst. Ser. S* 14 (2020), 2435–2453 [https://doi.org/10.3934/dcdss.2020401].

[25] B. Padhy, P. N. Samanta, S. K. Paikray and U. K. Misra, An EOQ model for items having fuzzy amelioration and deterioration, *Appl. Math. Inf. Sci.* 16 (2022) 353–360.

Chapter 7

A fuzzy inventory model with permissible delay in payments for non-instantaneous deteriorating items under inflation

Vipin Kumar, Amit Kumar, Chandra Shekhar, and C. B. Gupta

7.1 INTRODUCTION

Most researchers have developed their inventory models by assuming deterministic demand and deterioration of the product in order to minimize the total costs and predict optimal replenishment amounts and timings. But due to the fuzzy nature of the items, the demand and deterioration can increase or decrease, because this nature depends on many practical factors such as change in environmental conditions, availability of substitute products, and popularity of the product. Hence, the retailer can follow these models to order the items in less or more quantity and earlier or later respectively than the actual requirement and time. For the past few decades, the inventory problem for perishable goods has been studied extensively. Deterioration of goods means the decrease in the usefulness of goods, which can be due to many reasons such as evaporation, dryness, deterioration damage, and spoilage.

The first research of optimal ordering policies using a constant deterioration rate was conducted by Ghare and Schrader in 1963. This model was expanded by Cowart and Philippe (1973) with a two-parameter Weibull distribution over ten years later. The two-level permitted delay in payments fuzzy inventory model was discussed by N. Liu et al. (2010) as a way to increase demand. The two shops inventory model for deteriorating items with stock-dependent demand under inflation was explored by Singh et al. (2010). An inventory model was created by Chandra K. Jaggi et al. (2011) in a fuzzy environment with an allowable payment delay for non-immediately decaying goods. Chandra K. Jaggi et al. (2012) designed an inventory model with time-dependent demand and shortages in a fuzzy environment. In their inventory model, Vipin Kumar et al. (2013) include multivariate demand and parabolic holding costs with trade credit. Dutta et al. (2013) discussed their fuzzy inventory model's entirely backlogged shortages. D. Sharmila et al. (2015) examined an inventory model for decaying goods

DOI: 10.1201/9781003462422-7

that included shortages, exponential demand, and fuzzy. Fuzzy inventory model with time-dependent demand for degrading commodities, purposed by S. Kumar et al. (2015). S.R. Singh et al. (2016) introduce seasonal demand and demand based on available supply in their inventory model. Gopal Pathak et al. (2017) derived an inventory model in an inflationary setting. A supply chain fuzzy inventory model for degrading goods with price-dependent demand and no shortage was examined by Sujata Saha (2017). The demand is dependent on the selling price and the frequency of the advertisements under the conditions of allowable payment delays with changing demand and scarcity, according to the fuzzy inventory model developed by A. A. Shaikh et al. (2018). A fuzzy economic order quantity (EOQ) model without shortages and stock-dependent demand for decaying goods is proposed by S. K. Indrajitsingha et al. (2018). An inventory model with a time-varying supply chain system fuzzy environment was created by S. Shee et al. (2020). Fuzzy multi-item and multi-outlet inventory models for decaying items were discussed by Ajoy Kumar Maiti (2020). Vipin Kumar et al. (2020) assumed that the selling price and stock-dependent demand included trade credit. In order to boost sales, Boina Anil Kumar et al. (2020) created a two-warehouse inventory model by taking into account fuzzy demand and fuzzy deterioration. An EPQ for deteriorating goods with the effect of a price discount and demand that is stock-dependent was created by Sanjay Sharma et al. in 2021. An advance payment inventory model with deterioration and time-dependent demand was created by P. Supakar et al. in 2021. A reverse logistics inventory model of smart objects with carbon emissions in a fuzzy environment was examined by Subhash Kumar et al. in 2021. Smita Rani et al. (2022) developed a green supply chain fuzzy inventory model by applying the principle of reverse logistics, collecting and refurbishing used electronics products and selling them at a lower price than the market.

Keeping in view the important facts discussed in the above literature survey, we develop an inventory model in fuzzy environment which is applicable to non-instantaneous perishable items. Since uncertainty in the cost parameters used in any inventory system leads to a lot of impreciseness, fuzzy set theory has been used to handle it. In this model, we have considered triangular fuzzy numbers specifically for the cost parameters like ordering cost, holding cost, shortage cost, opportunity cost, deterioration cost, purchasing cost. The demand rate is time and selling price dependent. Shortages are allowed and partially backlogged. Permissible delays in payments and inflation also consider to create a more realistic inventory model. Finally, we discussed some numerical examples on each case of crisp and fuzzy inventory model to illustrate the model. After the Introduction section, Section 7.2 provides basic definitions related to the proposed model. Section 7.3 describes the notation and assumptions used during this study. Mathematical modeling and calculation of various components associated with the model are given in Section 7.4. In

Section 7.5, numerical solutions are given to illustrate the model. Finally, in the last section, the conclusions and future scope are given.

7.2 PRELIMINARIES

7.2.1 Definition

Fuzzy set
  Let X be a space of points (objects). A fuzzy set A in X is an object of the form $A=\{(x,\mu_A(x)):x\in X\}$ where $\mu_A:X\rightarrow[0,1]$ is called the membership function of the fuzzy set A

7.2.2 Definition

Triangular fuzzy numbers (TFNs)
  A triangular fuzzy number $A(a_1,a_2,a_3)$ is said to have membership function and defined as

$$\mu_A(x) = \begin{cases} \dfrac{x-a_1}{a_2-a_1}, & a_1 \leq x \leq a_2 \\[2mm] \dfrac{a_3-x}{a_3-a_2}, & a_1 \leq x \leq a_2 \\[2mm] 0, & \text{otherwise} \end{cases}$$

Defuzzification Method
  Let $(\tilde{a}_1,\tilde{a}_2,\tilde{a}_3)$ be triangular fuzzy numbers, its signed-distance formula is given by

$$\frac{a_1+2a_2+a_3}{4}$$

7.3 ASSUMPTIONS AND NOTATIONS

The following notation and assumptions are used throughout this chapter:

7.3.1 Assumption

  i. The demand rate is depending on selling price and time and defined as $D(p,t)=(a-bs)e^{\lambda kt}$, where $a>0, b>0$, where $0<k<1$ and $(\lambda>0)$.
  ii. Shortages are allowed and defined as $\beta(x)=e^{-\delta x}$, where $0<\delta<1$.
  iii. The rate of deterioration is constant.
  iv. The model is developed for non-instantaneous deteriorating items.
  v. The rate of replenishment is infinite.

7.3.2 Notation

A : The ordering cost.
$\tilde{A}$: Fuzzy ordering cost
θ : Rate of deterioration rate
$\tilde{\theta}$: Fuzzy rate of deterioration
C_1 : Holding cost
$\tilde{C}_1$: Fuzzy holding cost
C_2 : Shortage cost
$\tilde{C}_2$: Fuzzy shortage cost
C_3 : Opportunity cost
$\tilde{C}_3$: Fuzzy opportunity cost
p : Purchasing cost
$\tilde{p}$: Fuzzy purchasing cost
s : Selling price
I_p : Rate of interest on payable items amount.
I_e Rate of interest earned.
r : Rate of inflation.
$Q = I_0 + s$: The ordering quantity, where I_0 and S are maximum inventory and maximum backorder level, respectively
M : The trade credit period
t_d : Time from which spoilage begins.
t_1 : Time from which shortages start.
T replenishment cycle.
The inventory level at any time t during the interval $[0, t_d]$, $[t_d, t_1]$, and $[t_1, T]$ is
$I_1(t), I_2(t)$, and $I_3(t)$ respectively.
$TP(s, t_1, T)$ and $\widetilde{TP}(s, t_1, T)$ are the total profit function per unit time for crisp and fuzzy model, respectively, during the inventory cycle.

7.4 MATHEMATICAL MODEL

Entire inventory cycle is divided into three intervals, as shown in Figure 7.1. Therefore, in the first interval from 0 to t_d, inventory $I_1(t)$ is reduced only by demand, as depleted items are not immediately deteriorating. In the second interval from t_d to t_1, inventory decreases to zero level due to the combined effects of demand and deterioration. The last interval from t_1 to T shows a product shortage. The inventory at any point in time is determined by the first-order differential equation:

$$\frac{dI_1(t)}{dt} = -(a - bs)e^{\lambda kt}, \qquad\qquad 0 \le t \le t_d \qquad (7.1)$$

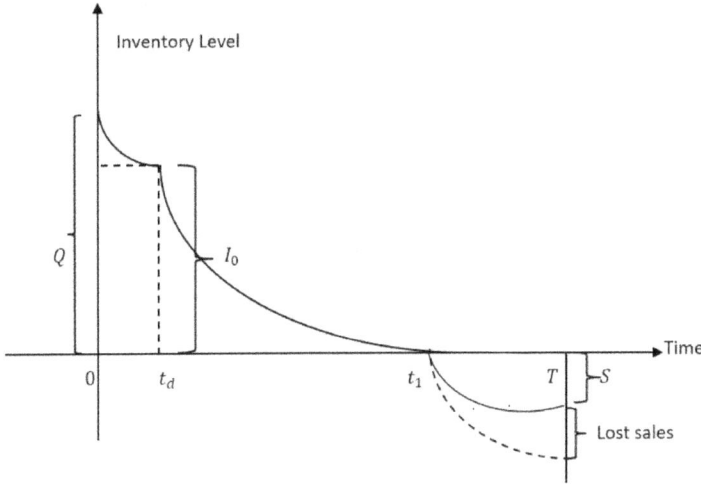

Figure 7.1 Graphical representation of model.

$$\frac{dI_2(t)}{dt} + \theta I_2(t) = -(a-bs)e^{\lambda kt}, \qquad\qquad t_d \leq t \leq t_1 \qquad\qquad (7.2)$$

$$\frac{dI_3(t)}{dt} = -(a-bs)e^{\lambda kt}e^{-\delta(T-t)}, \qquad\qquad t_1 \leq t \leq T \qquad\qquad (7.3)$$

under the boundary conditions

$$I_1(0) = I_0,\ I_2(t_1) = 0,\ \text{and}\ I_3(t_1) = 0.$$

On solving differential equations (7.1–7.3) by using boundary conditions, we get

$$I_1(t) = (a-bs)\frac{\left(1-e^{kt\lambda}\right)}{k\lambda} + I_0 \qquad\qquad 0 \leq t \leq t_d \qquad\qquad (7.4)$$

$$I_2(t) = \frac{(a-bs)}{\theta+k\lambda}\left(-e^{t(\theta+k\lambda)} + e^{(\theta+k\lambda)t_1}\right)e^{-t\theta} \qquad\qquad t_d \leq t \leq t_1 \qquad\qquad (7.5)$$

$$I_3(t) = \frac{(a-bs)}{\delta+k\lambda}\left(e^{(\delta+k\lambda)t_1} - e^{(\delta+k\lambda)t}\right)e^{-T\delta} \qquad\qquad t_1 \leq t \leq T. \qquad\qquad (7.6)$$

Considering continuity of $I_1(t_d) = I_2(t_d)$ by equations (7.4) and (7.5), we get

$$I_0 = (a-bs)\left[\frac{e^{t_d k\lambda}\theta}{k\lambda(\theta+k\lambda)} + \frac{e^{(t_1-t_d)\theta+k\lambda t_1}}{(\theta+k\lambda)} - \frac{1}{k\lambda}\right]. \qquad\qquad (7.7)$$

Substituting (7.7) in (7.4), we get

$$I_1(t) = (a-bs)\left[\frac{e^{t_d k\lambda}\theta}{k\lambda(\theta+k\lambda)} + \frac{e^{(t_1-t_d)\theta+k\lambda t_1}}{(\theta+k\lambda)} - e^{kt\lambda}\frac{1}{k\lambda}\right] \quad 0 \le t \le t_d, \quad (7.8)$$

The maximum backorders can be obtained by (7.6) as

$$S = -I_3(T) = \frac{(a-bs)}{\delta+k\lambda}\left(e^{T(\delta+k\lambda)} - e^{(\delta+k\lambda)t_1}\right)e^{-T\delta}. \quad (7.9)$$

Per cycle total ordering quantity

$$Q = I_0 + S = (a-bs)\left[\frac{e^t d^{k\lambda}\theta}{k\lambda(\theta+k\lambda)} + \frac{e^{(t_1-t_d)\theta+k\lambda t_1}}{(\theta+k\lambda)} - \frac{e^{kt\lambda}}{k\lambda}\right.$$

$$\left. + \frac{1}{\delta+k\lambda}\left(e^{k\lambda T} - e^{(\delta+k\lambda)t_1 - T\delta}\right)\right] \quad (7.10)$$

The following are the component of total profit function

I. Ordering Cost:
A
II. Holding Cost

$$HC = C_1\left[\int_0^{t_d} I_1(t)e^{-rt}dt + \int_{t_d}^{t_1} I_2(t)e^{-rt}dt\right]$$

$$= \frac{C_1(a-bs)}{\theta+k\lambda}\left(\frac{e^{k\lambda t_1}\left(e^{-rt_1} - e^{-rt_d-t_d\theta+\theta t_1}\right)}{r+\theta} + \frac{e^{-rt_d+kt_d\lambda} - e^{-rt_1+k\lambda t_1}}{r-k\lambda}\right.$$

$$+ \frac{1}{k\lambda}\left(\frac{\theta+k\lambda}{-r+k\lambda} + \frac{e^{kt_d\lambda}\theta + e^{-t_d\theta+\theta t_1+k\lambda t_1}k\lambda}{r}\right.$$

$$\left.\left. + e^{-rt_1}\left(\frac{e^{k\lambda t_1}(\theta+k\lambda)}{r-k\lambda} - \frac{e^{kt_d\lambda}\theta + e^{-t_d\theta+\theta t_1+k\lambda t_1}k\lambda}{r}\right)\right)\right)$$

III. The shortages cost

$$SC = C_2\int_{t_1}^{T} -I_3(t)e^{-rt}dt$$

$$= C_2\frac{(a-bs)}{\delta+k\lambda}\left(\frac{e^{-r(T+t_1)+(\delta+k\lambda)t_d}\left(e^{rt_1} - e^{rT}\right)}{r} + \frac{e^{T(\delta+k\lambda-r)} - e^{t_1(\delta+k\lambda-r)}}{-r+\delta+k\lambda}\right)e^{-T\delta}$$

IV. The opportunity cost

$$OC = C_3\int_{t_1}^{T} D(p,t)(1-\beta(T-t))e^{-rt}dt$$

$$= C_3 \frac{(a-bs)}{(r-k\lambda)(r+\delta-k\lambda)}$$

$$\left(e^{-rt_1+kt_1\lambda}\left(\delta+(r-k\lambda)\left(1-e^{\delta(T-t_1)}\right)\right)-e^{-rT+kT\lambda}\delta\right)$$

V. The present value of purchase cost

$$PC = c \times Q$$

$$= c(a-bs)\left[\frac{e^{t_d k\lambda}\theta}{k\lambda(\theta+k\lambda)} + \frac{e^{(t_1-t_d)\theta+k\lambda t_1}}{(\theta+k\lambda)} - \frac{e^{kt\lambda}}{k\lambda}\right.$$

$$\left. + \frac{1}{\delta+k\lambda}\left(e^{k\lambda T}-e^{(\delta+k\lambda)t_1-T\delta}\right)\right]$$

VI. The present value of sales revenue

$$SR = s\left[\int_0^{t_1} D(s,t) e^{-rt}dt + Se^{-rT}\right]$$

$$= s(a-bs)\left(\frac{1-e^{-r\eta+k\eta\lambda}}{r-k\lambda} + \frac{e^{-T(r+\delta)}\left(-e^{T(\delta+k\lambda)}+e^{(\delta+k\lambda)v}\right)}{\delta+k\lambda}\right)$$

Case I $0 < M \le t_d$

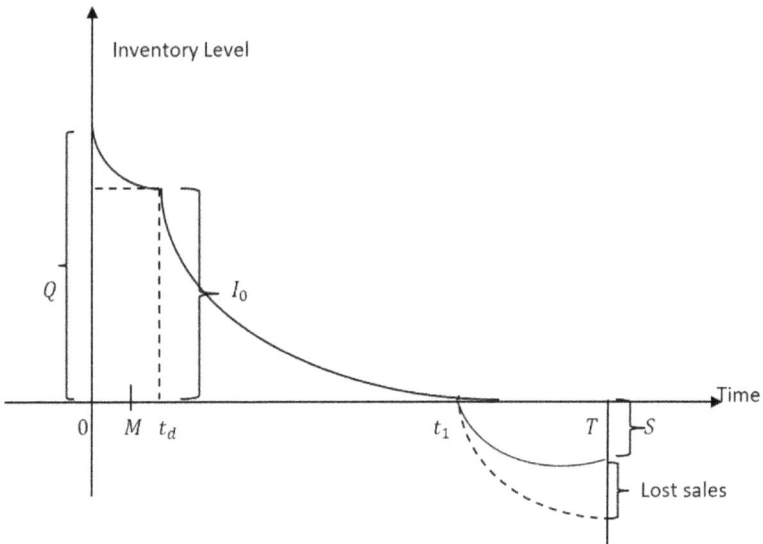

Figure 7.2 Case I: $0 \le M \le t_d$.

Interest payable by the retailer

$$IP_1 = pI_p \int_M^{t_d} I_1(t) e^{-rt} dt + pI_p \int_{t_d}^{t_1} I_2(t) e^{-rt} dt$$

$$= \frac{pI_p(a-bs)}{\theta + k\lambda} \left(\frac{e^{k\lambda t_d}\left(e^{-rt_d} - e^{-rt_1 - t_1\theta + \theta t_d}\right)}{r+\theta} + \frac{e^{-rt_1 + kt_1\lambda} - e^{-rt_d + k\lambda t_d}}{r - k\lambda} \right.$$

$$+ \frac{e^{-rt_d}}{k\lambda}\left(\frac{e^{k\lambda t_d}(\theta + k\lambda)}{r - k\lambda} - \frac{e^{kt_1\lambda}\theta + e^{-t_1\theta + \theta t_d + k\lambda t_d}k\lambda}{r} \right)$$

$$\left. + \frac{e^{-Mr}}{k\lambda}\left(\frac{e^{kM\lambda}(\theta + k\lambda)}{-r + k\lambda} + \frac{e^{kt_1\lambda}\theta + e^{-t_1\theta + \theta t_d + k\lambda t_d}k\lambda}{r} \right) \right)$$

Interest earned by the retailer

$$IE_1 = sI_e \int_0^M D(s,t) t e^{-rt} dt$$

$$= \frac{s(a-bs)I_e}{(r-k\lambda)^2}\left(1 + e^{kM\lambda - Mr}\left(-1 - Mr + kM\lambda \right) \right)$$

Case II $t_d < M \le t_1$

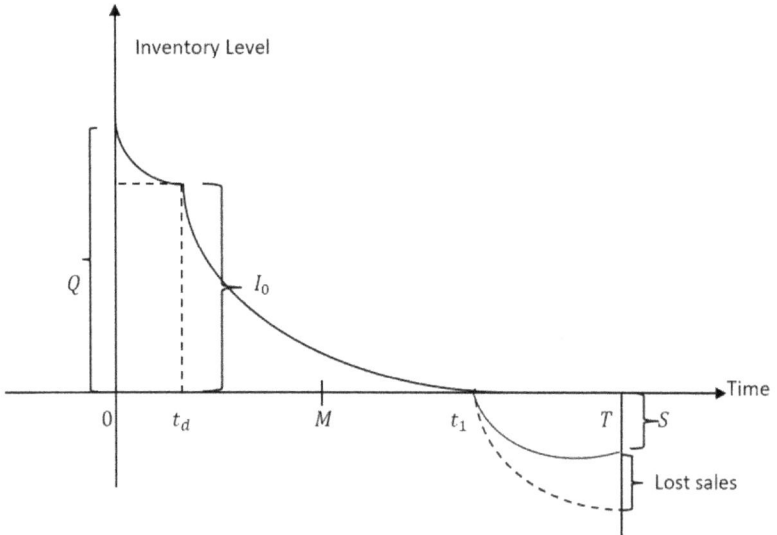

Figure 7.3 Case II: $t_d \le M \le t_1$.

VII. Interest payable by the retailer

$$IP_2 = pI_p \int_M^{t_1} I_2(t) e^{-rt} dt$$

$$= \frac{pI_p(a-bs)}{\theta+k\lambda} \left(\frac{e^{-(M+t_1)(r+\theta)+(\theta+k\lambda)t_d} \left(-e^{M(r+\theta)} + e^{t_1(r+\theta)} \right)}{r+\theta} \right.$$

$$\left. \frac{e^{-Mr+kM\lambda} - e^{-rt_1+kt_1\lambda}}{r-k\lambda} \right)$$

VIII. Interest earned by the retailer

$$IE_2 = sI_e \int_0^M D(s,t) te^{-rt} dt$$

$$= \frac{s(a-bs)I_e}{(r-k\lambda)^2} \left(1 + e^{kM\lambda - Mr} \left(-1 - Mr + kM\lambda \right) \right)$$

Case III $t_1 < M \le T$

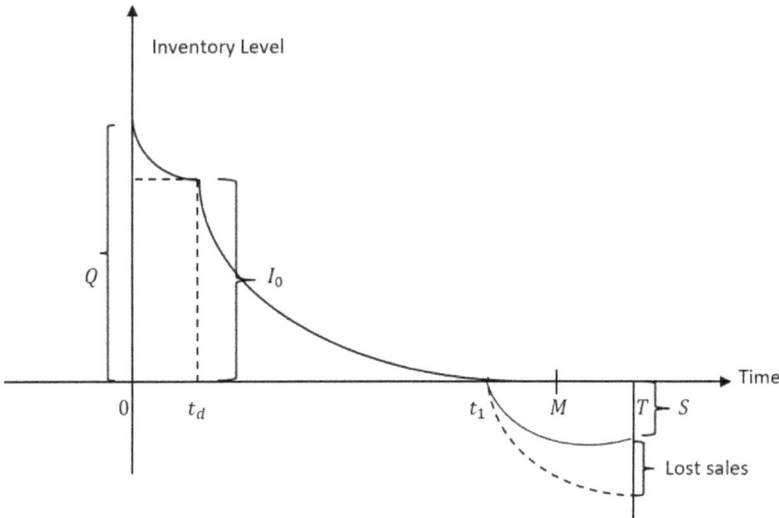

Figure 7.4 Case III: $t_1 \le M \le T$.

Interest payable by the retailer

$$IP_3 = 0$$

Interest earned by the retailer

$$IE_3 = sI_e \left[\int_0^{t_1} D(s,t) t e^{-rt} dt + (M - t_1) \int_0^{t_1} D(s,t) e^{-rt} dt \right]$$

$$= sI_e (a - bs) \left(\frac{e^{-rt_1} \left(e^{rt_1} - e^{kt_1\lambda} \right)(M - t_1)}{r - k\lambda} \right.$$

$$\left. + \frac{e^{-rt_1} \left(e^{rt_1} + e^{kt_1\lambda} \left(-1 - rt_1 + kt_1\lambda \right) \right)}{(r - k\lambda)^2} \right)$$

Crisp optimization model
The total profit function is

$$TP(s, t_1, T) = \begin{cases} TP_1(s, t_1, T) \\ TP_2(s, t_1, T) \\ TP_3(s, t_1, T) \end{cases}$$

where $TP_1(s, t_1, T) = \dfrac{SR - A - HC - SC - OC - PC - IP_1 + IE_1}{T}$,

$$TP_2(s, t_1, T) = \frac{SR - A - HC - SC - OC - PC - IP_2 + IE_2}{T},$$

and $TP_3(s, t_1, T) = \dfrac{SR - A - HC - SC - OC - PC - IP_3 + IE_3}{T}$

7.4.1 Fuzzy optimization model

We can change only a few parameters like A, θ, C_1, C_2, C_3, and p within a certain range, because it is not possible to accurately describe all the parameters due to uncertainty.

Let $\tilde{A}$, $\tilde{\theta}$, $\widetilde{C_1}$, $\widetilde{C_2}$ $\widetilde{C_3}$, and $\tilde{p}$ be triangular fuzzy numbers.

Let us assume that parameters $\tilde{A}$, $\tilde{s}$, $\tilde{\theta}$, $\widetilde{C_1}$, $\widetilde{C_2}$, $\widetilde{C_3}$, and $\tilde{p}$ may change within some limits.

$$\tilde{A} = (A - \Delta_1, A, A + \Delta_2), \quad \text{where } \Delta_1, \Delta_2 > 0 \quad \text{and } 0 < \Delta_1 < A$$

$$\tilde{\theta} = (\theta - \Delta_3, \theta, \theta + \Delta_4), \quad \text{where } \Delta_3, \Delta_4 > 0 \quad \text{and } 0 < \Delta_3 < \theta$$

$$\widetilde{C_1} = (C_1 - \Delta_5, C_1, C_1 + \Delta_6), \quad \text{where } \Delta_5, \Delta_6 > 0 \quad \text{and } 0 < \Delta_5 < C_1$$

$$\widetilde{C}_2 = \left(C_2 - \Delta_7, C_2, C_2 + \Delta_8\right), \quad \text{where } \Delta_7, \Delta_8 > 0 \qquad \text{and } 0 < \Delta_7 < C_2$$

$$\widetilde{C}_3 = \left(C_3 - \Delta_9, C_3, C_3 + \Delta_{10}\right), \quad \text{where } \Delta_9, \Delta_{10} > 0 \qquad \text{and } 0 < \Delta_9 < C_3$$

$$\tilde{p} = \left(p - \Delta_{11}, p, p + \Delta_{12}\right), \qquad \text{where } \Delta_{11}, \Delta_{12} > 0 \qquad \text{and } 0 < \Delta_{11} < p.$$

By signed-distance method

$$\tilde{A} = A + \frac{1}{4}\left(\Delta_2 - \Delta_1\right)$$

$$\tilde{\theta} = \theta + \frac{1}{4}\left(\Delta_4 - \Delta_3\right)$$

$$\widetilde{C}_1 = C_1 + \frac{1}{4}\left(\Delta_6 - \Delta_5\right)$$

$$\widetilde{C}_2 = C_2 + \frac{1}{4}\left(\Delta_8 - \Delta_7\right)$$

$$\widetilde{C}_3 = C_3 + \frac{1}{4}\left(\Delta_{10} - \Delta_9\right)$$

$$\tilde{p} = p + \frac{1}{4}\left(\Delta_{12} - \Delta_{11}\right)$$

Now defuzzified total cost per unit time by signed-distance method is given by

$$\widetilde{TP}(s, t_1, T) = \begin{cases} TP_1(s, t_1, T) \\ TP_2(s, t_1, T) \\ TP_3(s, t_1, T) \end{cases}$$

where

$$TP_i(s, t_1, T) = \frac{1}{4}\left[\widetilde{TP_iF_1}(s, t_1, T) + \widetilde{2TP_iF_2}(s, t_1, T) + \widetilde{TP_iF_3}(s, t_1, T)\right],$$

$$i = 1, 2, 3.$$

7.5 NUMERICAL EXAMPLES

7.5.1 Crisp numerical example

Example 1 $\left[0 < M \le t_d\right]$

Consider the following numerical values:

$A = \$250 \ /order, C_1 = \$1 \ /unit \ , C_2 = \$5 \ /unit,$

$C_3 = \$25 \ /unit, p = \$20 \ /unit, Ie = 0.12, Ip = 0.15, r = 0.001,$

$\theta = 0.08, a = 200, b = 4, \delta = 0.1, \lambda = 0.98, t_d = 0.42, \text{ and } M = 0.12$

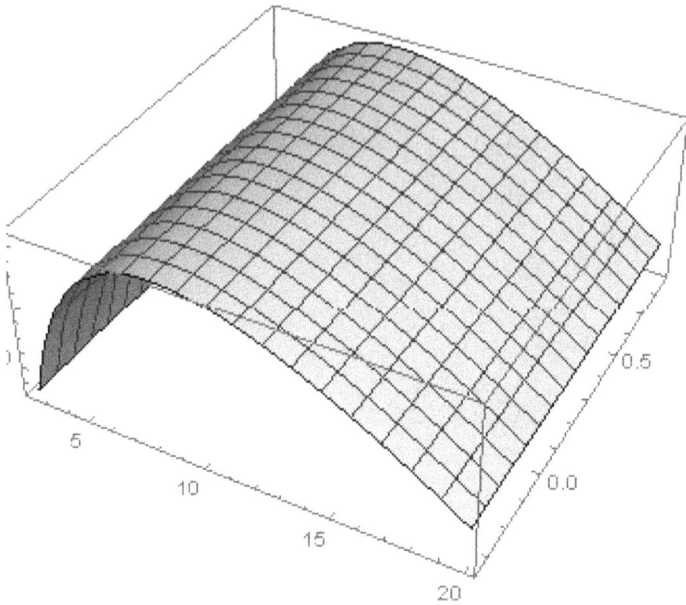

Figure 7.5 Total profit per unit time.

Optimal solution is

$$s^* = 448.9127, t_1^* = 0.717, T^* = 0.911, TP^* = 56118.21, \text{ and } Q^* = 21.0271$$

Example 2 $[t_d < M \le t_1]$

Consider the following numerical values:

$A = \$250 \, /order, C_1 = \$1 \, /unit, C_2 = \$5 \, /unit, C_3 = \$25 \, /unit,$
$p = \$20 \, /unit, \ Ie = 0.12, Ip = 0.15, r = 0.001,$
$\theta = 0.08, a = 200, b = 4, \delta = 0.1, \lambda = 0.98 t_d = 0.42 \text{ ,and } M = 0.53 \, .$

Optimal solution is

$$s^* = 498.2127, t_1^* = 0.693, T^* = 0.709, TP^* = 49871.12, \text{ and } Q^* = 19.9856$$

Example 3 $[t_1 < M \le T]$

Consider the following numerical values:

$A = \$250 \, /order, C_1 = \$1 \, /unit, C_2 = \$5 \, /unit, C_3 = \$25 \, /unit,$
$p = \$20 \, /unit, Ie = 0.12, Ip = 0.15, r = 0.001, \theta = 0.08,$
$a = 200, b = 4, \delta = 0.1, \lambda = 0.98 t_d = 0.42, \text{ and } M = 0.76$

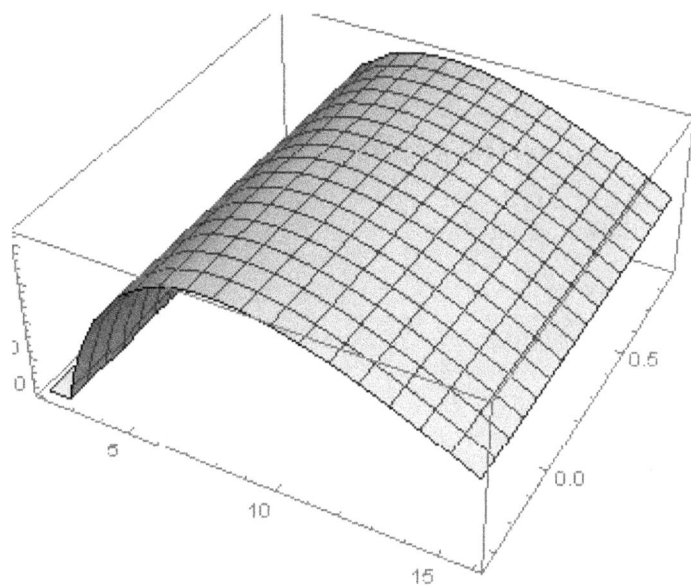

Figure 7.6 Total profit per unit time.

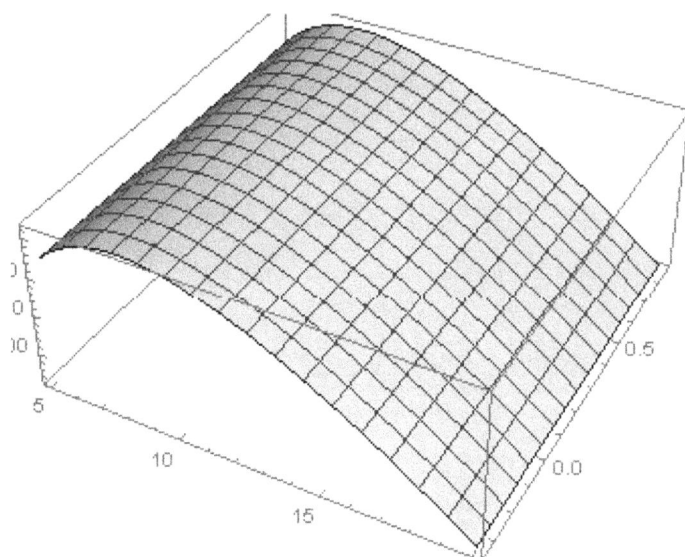

Figure 7.7 Total profit per unit time.

Optimal solution is

$$s^* = 501.7653, t_1^* = 0.602, T^* = 0.873, TP^* = 48731.23, \text{ and } Q^* = 20.5463$$

7.5.2 Fuzzy numerical example

Example 4 $[0 < M \le t_d]$

Consider the following numerical values:

$$\tilde{A} = (200, 250, 300), \; \tilde{C_1} = (0.8, 1, 1.2), \; \tilde{C_2} = (4, 5, 6), \; \tilde{C_3} = (20, 25, 30),$$
$$\tilde{\theta} = (0.64, 0.08, 0.96), \; \tilde{p} = (16, 20, 24),$$

$$Ie = 0.12, Ip = 0.15, r = 0.001, \theta = 0.08, a = 200, b = 4,$$
$$\delta = 0.1, \lambda = 0.98, t_d = 0.42, \text{ and } M = 0.12$$

Optimal solution is

$$s^* = 407.2191, t_1^* = 0.514, \; T^* = 0.836, \; TP^* = 42423.91, \text{ and } Q^* = 22.05101.$$

Example 5 $[t_d < M \le t_1]$

Consider the following numerical values

$$\tilde{A} = (200, 250, 300), \; \tilde{C_1} = (0.8, 1, 1.2), \; \tilde{C_2} = (4, 5, 6), \; \tilde{C_3} = (20, 25, 30),$$
$$\tilde{\theta} = (0.64, 0.08, 0.96) \; \tilde{p} = (16, 20, 24),$$

$$Ie = 0.12, Ip = 0.15, r = 0.001, \theta = 0.08, a = 200, b = 4,$$
$$\delta = 0.1, \lambda = 0.98, t_d = 0.42, \text{ and } M = 0.53$$

Optimal solution is

$$s^* = 417.9331, t_1^* = 0.576, T^* = 0.911, TP^* = 56893.98, \text{and } Q^* = 22.2137$$

Example 6 $[t_1 < M \le T]$

Consider the following numerical values:

$$\tilde{A} = (200, 250, 300), \; \tilde{C_1} = (0.8, 1, 1.2), \; \tilde{C_2} = (4, 5, 6), \; \tilde{C_3} = (20, 25, 30),$$
$$\tilde{\theta} = (0.64, 0.08, 0.96), \; \tilde{p} = (16, 20, 24),$$

$$Ie = 0.12, Ip = 0.15, r = 0.001, \theta = 0.08, a = 200, b = 4,$$
$$\delta = 0.1, \lambda = 0.98, t_d = 0.42, \text{ and } M = 0.76$$

Optimal solution is

$$s^* = 487.3351, t_1^* = 0.599, \; T^* = 0.773, TP^* = 47009.04, \text{ and } Q^* = 20.0041$$

7.6 CONCLUSION

In this study, we developed a fuzzy inventory model for non-instantaneous perishable goods, where demand is dependent on selling price and time. In addition, there is applied a trade credit policy with inflation. Here, fuzzy set theory is applied to deal with situations of inaccurate data and uncertainty in the market. Therefore, we described the order cost, inventory holding cost, deterioration cost, purchase cost, shortage cost, and opportunity cost with triangular fuzzy numbers, and defuzzified the total profit using the GMRM method. This chapter calculates the optimal selling price, optimal start point of stockout, optimal order quantity, and optimal replenishment cycle time to maximize overall profit. Numerical examples for each case are given to illustrate the model. This model can also be extended with variable deteriorating cost and inventory-dependent demand under cloudy fuzzy theory.

REFERENCES

Anil Kumar, Boina, S. K. Paikray and Hemen Dutta (2020) Cost optimization model for items having fuzzy demand and deterioration with two-warehouse facility under the trade credit financing [J]. *AIMS Mathematics*, 5(2), 1603–1620. doi:10.3934/math.2020109.

Covert, R. P. and G. C. Philip (1973) An EOQ model for item with Weibull distribution deterioration. *AIIE Trans*, 5, 323–326.

Dutta, Debashis and Pavan Kumar (2013) Fuzzy inventory model for deteriorating items with shortages under fully backlogged condition. *International Journal of Soft Computing and Engineering (IJSCE)*, 3(2), May, 393–398. ISSN: 2231-2307.

Ghare, P. M. and G. F. Schrader (1963) A model for an exponential decaying inventory. *Journal of Industrial Engineering*, 14, 238–243.

Indrajit Singha, S. K., P. N. Samanta and U. K. Misra (2018) A fuzzy inventory model for deteriorating items with stock dependent demand rate. *International Journal of Logistics Systems and Management*, 30(4), 538–555.

Jaggi, Chandra K., Anuj Sharma and Mandeep Mittal (2011) A fuzzy inventory model for non- instantaneous deteriorating items under partial backlogging and permissible delay in payments. *International Journal of Inventory Control and Management Special Issue on International Conference on Applied Mathematics & Statistics*, December, 167–200.

Jaggi, Chandra K., Sarla Pareek, Anuj Sharma and A. Nidhi (2012) Fuzzy inventory model for deteriorating items with time-varying demand and shortages. *American Journal of Operational Research*, 2(6), 81–92.

Kumar, S. and U. Rajput (2015) Fuzzy inventory model for deteriorating items with time dependent demand and partial backlogging. *Applied Mathematics*, 6, 496–509. doi:10.4236/am.2015.63047.

Kumar, Subhash, Biswajit Sarkar and Ashok Kumar (2021) Fuzzy reverse logistics inventory model of smart items with two warehouses of a retailer considering carbon emissions. *RAIRO—Operations Research*, 55(4), 2285–2307.

Kumar, Vipin, Gopal Pathak and C. B. Gupta (2013) A deterministic inventory model for deteriorating items with selling price dependent demand and parabolic time varying holding cost under trade credit. *International Journal of Soft Computing and Engineering (IJSCE)*, 3(4), September, 33–37. ISSN: 2231-2307.

Kumar, Vipin, Anupama Sharma, Amit Kumar and C. B. Gupta (2020) An inventory model for deteriorating items with multivariate demand and trade credit. *Advances in Mathematics: Scientific Journal* 9(9), 7501–7514. doi:10.37418/amsj.9.9.97.

Liu, N. and J. Hu (2010) Fuzzy inventory model for deteriorating items under two-level permissible delay in payments. *2010 International Conference on E-Product E-Service and E-Entertainment*, 1–4.

Maiti, Ajoy Kumar (2020) Multi-item fuzzy inventory model for deteriorating items in multi-outlet under single management. *Journal of Management Analytics*, 7(1), 44–68. doi: 10.1080/23270012.2019.1699873.

Pathak, Gopal, Vipin Kumar and C. B. Gupta (2017) A cost minimization inventory model for deteriorating products and partial backlogging under inflationary environment. *Global Journal of Pure and Applied Mathematics*, 13(9), 5977–5995. ISSN 0973-1768.

Rani, Smita, Rashid Ali and Anchal Agarwal (2022) Fuzzy inventory model for new and refurbished deteriorating items with cannibalisation in green supply chain. *International Journal of Systems Science: Operations & Logistics*, 9(1), 22–38. doi: 10.1080/23302674.2020.1803434.

Saha, Sujata (2017) Fuzzy inventory model for deteriorating items in a supply chain system with price dependent demand and without backorder. *American Journal of Engineering Research (AJER)*, 6(6), 183–187.

Shaikh, A. A., A. K. Bhunia, L. E. Cárdenas-Barrón, et al. (2018) A fuzzy inventory model for a deteriorating item with variable demand, permissible delay in payments and partial backlogging with shortage. *International Journal of Fuzzy Systems*, 20, 1606–1623. doi:10.1007/s40815-018-0466-7.

Sharma, Sanjay, Vipin Kumar and Anand Tyagi (2021) A production inventory model for deteriorating items with effect of price discount under the stock dependent demand. *RT&A, Special Issue No 2(64)*, 16, November, 213–224.

Sharmila, D. and R. Uthayakumar (2015) Inventory model for deteriorating items involving fuzzy with shortages and exponential demand. *International Journal of Supply and Operations Management*, 2(3), November, 888–904.

Shee, Srabani and Tripti Chakrabarti (2020) Fuzzy inventory model for deteriorating items in a supply chain system with time dependent demand rate. *International Journal of Engineering Applied Sciences and Technology*, 5(1), 558–569, ISSN No. 2455-2143.

Singh, S. R., N. Kumar and R. Kumari (2010) An inventory model for deteriorating items with shortages and stock-dependent demand under inflation for two-shops under one management. *OPSEARCH*, 47, 311–329. doi:10.1007/s12597-010-0026-x.

Singh, S. R., Mohit Rastogi and Shilpy Tayal (2016) An inventory model for deteriorating items having seasonal and stock-dependent demand with allowable shortages. *Proceedings of Fifth International Conference on Soft Computing for Problem Solving. Advances in Intelligent Systems and Computing*, vol. 437, 501–513. doi:10.1007/978-981-10-0451-3_46.

Supakar, P., S. K. Mahato and P. Pal (2021) Intuitionistic fuzzy inventory model with deterioration incorporating advance payment with time-dependent demand. *International Journal of Applied and Computational Mathematics*, 7, 228. doi:10.1007/s40819-021-01149-5.

Chapter 8

Single- and multi-objective optimization problems with pentagonal fuzzy number

Anuradha Sahoo, Jayanta Kumar Dash, and Ruchika Moharana

8.1 INTRODUCTION

An optimization model is a key factor in several fields in handling many real-life problems. Many optimization techniques have been developed by many researchers and implemented in different areas of study to obtain the optimal solutions which optimize a single objective or multi-objectives. A multi-objective optimization problem (MOOP) has multiple objectives, and each of the objectives either minimizes or maximizes. An MOOP has a set of different objective functions subject to some constraints which are to be minimized or maximized. The constraints may be equality or inequality. The model may be a linear or nonlinear optimization problem.

In the real world, problems are associated with some set of uncertainties. And people used to believe that the probability theory could tackle the different types of uncertainties in this world. But Zadeh [1] took the concept of the fuzzy set to represent the different types of uncertainties that couldn't be tackled using the probability theory. Zimmermann [2] extended the concept of fuzzy set theory.

Yao and Wu [3] used the decomposition principle and crisp ranking system to defuzzify the fuzzy numbers. Yao and Chiang [4] proposed different defuzzification techniques like the maxima and signed-distance methods in a fuzzy model and compared the solution. Aziz Syed [5] used the signed-distance method for defuzzification in an inventory model consisting of fuzzy variables. Chen and Chang [6] discussed the optimization inventory model with irreparable defective products in a fuzzy sense.

Miettinen [7] discussed the global criterion method to transform multi-problem optimization into a single problem optimization. Das and Dennis [8] discussed the weighted sum method, where all objectives are combined into one objective by using the weighted functions. Marler and Arora [9] discussed the Pareto optimal solutions by using the weighted sum method for MOOP. Ramli and Jaaman [10] proposed an extended classical Markowitz's mean-variance model where the fuzzy pentagonal number represents the returns. Das and Chakraborty [11] disclosed the concept of the pentagonal neutrosophic approach to solve linear programming problems. Alharbi and Khalifa [12] presented a simple technique for the flow-shop scheduling

DOI: 10.1201/9781003462422-8

problem under a fuzzy environment. They considered fuzzy pentagonal numbers for the processing time of jobs.

This chapter is structured as follows. Section 8.2 gives some basic preliminaries about the fuzzy set. Section 8.3 presents the mathematical optimization model for both crisp and fuzzy sets. In Section 8.4, the methodology for the optimization models is given. Section 8.5 gives a numerical example to verify the said model. Finally, the conclusion is given in Section 8.6.

8.2 BASIC PRELIMINARIES

To discuss a fuzzy model, we need to know the following definitions.

8.2.1 Fuzzy set

A fuzzy $\tilde{A}$ set is defined on a universal set X, which is written as:

$\tilde{A} = \{(x, \mu_{\tilde{A}}(x)) : x \in X\}$, where $\mu_A(x)$: membership function of x in $\tilde{A}$.

8.2.2 Fuzzy number

Fuzzy set $\tilde{A}$ is called a fuzzy number if the following conditions are satisfied:

- $\tilde{A}$ is a convex fuzzy set.
- $\tilde{A}$ is a normalized fuzzy set.
- Its membership function is piecewise continuous.

8.2.3 α-cut of fuzzy number

$\alpha-cut$ of $\tilde{A}$ on X is defined as

$\tilde{A} = \{x \in X : \mu_{\tilde{A}}(x) \geq \alpha\}, 0 \leq \alpha \leq 1$

8.2.4 Pentagonal fuzzy number

Let $\tilde{A} = (a, b, c, d, e; w)$, $a < b < c < d < e$, be a fuzzy pentagonal number represented with membership function $\mu_{\tilde{A}}(x)$ as:

$$\mu_{\tilde{A}}(x) = \begin{cases} 0, & x < a \\ w\dfrac{x-a}{b-a}, & a \leq x \leq b \\ 1-(1-w)\dfrac{x-b}{c-b}, & b \leq x \leq c \\ 1, & x = c \\ 1-(1-w)\dfrac{d-x}{d-c}, & c \leq x \leq d \\ w\dfrac{e-x}{e-d}, & d \leq x \leq e \\ 0, & x > e \end{cases}$$

where $0 \leq w \leq 1$

8.2.4.1 α-cut of pentagonal fuzzy number

The α-cuts of $\tilde{A} = (a,b,c,d,e)$ are

$$A(\alpha) = [A_L(\alpha), A_R(\alpha)]$$

where $A_L(\alpha) = \left[a + \dfrac{(b-a)\alpha}{w}, e - \dfrac{(e-d)\alpha}{w} \right]$, For $\alpha \in [0,w]$

$$A_R(\alpha) = \left[b + \dfrac{(c-b)(1-\alpha)}{1-w}, d - \dfrac{(d-c)(1-\alpha)}{1-w} \right], \text{ For } \alpha \in [w,1]$$

8.2.5 Signed-distance method

We consider a family F of fuzzy numbers on $X = (-\infty, \infty)$. Let $\tilde{A} \epsilon F$ be a fuzzy set on X, such as $\tilde{A} = \{(x, \mu_{\tilde{A}}(x)) : x \in X\}$, where $\mu_{\tilde{A}}(x)$ maps X to the closed interval $[0,1]$.

Let $A_L(\alpha)$ and $A_R(\alpha)$ be the left and right α-cuts of the fuzzy set $\tilde{A}$, where $0 \le \alpha \le 1$. We assume that $A_L(\alpha)$ and $A_R(\alpha)$ exist and are integrals for $\alpha \in [0,1]$. Moreover, the fuzzy set $\tilde{A}$ can be represented using its α-cuts as:

$$\tilde{A} = \bigcup_{0 \le \alpha \le 1} [A_L(\alpha), A_R(\alpha); \alpha]$$

Now the signed distance of $\tilde{A}$ from the fuzzy origin $\tilde{O}$ is:

$$d(\tilde{A}, \tilde{0}) = \frac{1}{2} \int_0^1 [A_L(\alpha) + A_R(\alpha)] d\alpha$$

8.2.5.1 Signed distance of a fuzzy pentagonal number

The signed distance of a fuzzy pentagonal number $\tilde{A} = (a,b,c,d,e)$ can be calculated as

$$d(\tilde{A}, \tilde{0}) = \frac{1}{2} \int_0^1 [A_L(\alpha) + A_R(\alpha)] d\alpha$$

$$= \frac{1}{2} \left[aw + \frac{(b-a)w}{2} + ew - \frac{(e-d)w}{2} + b(1-w) \right.$$

$$\left. + \frac{(c-b)(1-w)}{2} + d(1-w) - \frac{(d-c)(1-w)}{2} \right]$$

8.3 MATHEMATICAL MODEL

Here, we described different optimization models in both crisp and fuzzy environments. The said model may be linear or nonlinear, which consists of n number of variables and m number of constraints. The constraints may be equality or inequality.

8.3.1 Single-objective optimization model

The following subsections describe the single-objective optimization model in both a crisp and fuzzy sense.

8.3.1.1 Crisp single-objective optimization model (CSOOM)

The mathematical model of CSOOM with n number of variables and m number of constraints is written as follows:

$$Min \quad c_1 x_1 + c_2 x_2 + ... + c_n x_n$$

$$s.t. \quad a_{11} x_1 + a_{12} x_2 + ... + a_{1n} x_n \leq / = / \geq b_1$$

$$\text{CSOOM:} \quad a_{21} x_1 + a_{22} x_2 + ... + a_{2n} x_n \leq / = / \geq b_2$$

$$\cdots\cdots\cdots\cdots\cdots\cdots\cdots\cdots\cdots\cdots\cdots\cdots\cdots$$

$$a_{m1} x_1 + a_{m2} x_2 + ... + a_{mn} x_n \leq / = / \geq b_m$$

$$x_1, x_2, ... x_n \geq 0$$

In the above model, all the coefficients of objective function and constraints are deterministic in nature.

8.3.1.2 Fuzzy single-objective optimization model (FSOOM)

Here, the coefficient of the objective function and constraints are fuzzy in nature, and we considered all the coefficients and the right-hand side of the constraints for the FSOOM as fuzzy pentagonal numbers. The mathematical model of FSOOM with n number of variables and m number of constraints is written as follows:

$$Min \quad \tilde{c}_1 x_1 + \tilde{c}_2 x_2 + ... + \tilde{c}_n x_n$$

$$s.t. \quad \tilde{a}_{11} x_1 + \tilde{a}_{12} x_2 + ... + \tilde{a}_{1n} x_n \leq / = / \geq \tilde{b}_1$$

$$\text{FSOOM:} \quad \tilde{a}_{21} x_1 + \tilde{a}_{22} x_2 + ... + \tilde{a}_{2n} x_n \leq / = / \geq \tilde{b}_2$$

$$\cdots\cdots\cdots\cdots\cdots\cdots\cdots\cdots\cdots\cdots\cdots\cdots$$

$$\tilde{a}_{m1}x_1 + \tilde{a}_{m2}x_2 + \ldots + \tilde{a}_{mn}x_n \leq / = / \geq \tilde{b}_m$$

$$x_1, x_2, \ldots x_n \geq 0$$

8.3.2 Multi-objective optimization model

The multi-objective optimization problem can be defined as the collection of objective functions $z_1(x), z_2(x), \ldots z_n(x)$, which is to be optimized and subject to some constraints. Here, we have taken the described model with m number of constraints and n number of variables. The following two subsections describe the multi-objective optimization model in the crisp and fuzzy sense.

8.3.2.1 Crisp multi-objective optimization model (CMOOP)

The mathematical model of CMOOP with n number of variables and m number of constraints is written as follows:

$$Min \qquad z_1(x), z_2(x), \ldots, z_n(x)$$

$$s.t. \qquad a_{11}x_1 + a_{12}x_2 + \ldots + a_{1n}x_n \leq / = / \geq b_1$$

CMOOM: $\qquad a_{21}x_1 + a_{22}x_2 + \ldots + a_{2n}x_n \leq / = / \geq b_2$

$$\cdots\cdots\cdots\cdots\cdots\cdots\cdots\cdots\cdots\cdots\cdots\cdots$$

$$a_{m1}x_1 + a_{m2}x_2 + \ldots + a_{mn}x_n \leq / = / \geq b_m$$

$$x_1, x_2, \ldots x_n \geq 0$$

8.3.2.2 Fuzzy multi-objective optimization model (FMOOM)

The objective function and constraint coefficients in the FMOOM are considered fuzzy pentagonal numbers. The mathematical model of FMOOM with n number of variables and m number of constraints is written as follows:

$$Min \qquad \tilde{z}_1(x), \tilde{z}_2(x), \ldots \tilde{z}_n(x)$$

$$s.t. \qquad \tilde{a}_{11}x_1 + \tilde{a}_{12}x_2 + \ldots + \tilde{a}_{1n}x_n \leq / = / \geq \tilde{b}_1$$

FMOOM: $\qquad \tilde{a}_{21}x_1 + \tilde{a}_{22}x_2 + \ldots + \tilde{a}_{2n}x_n \leq / = / \geq \tilde{b}_2$

$$\cdots\cdots\cdots\cdots\cdots\cdots\cdots\cdots\cdots\cdots\cdots\cdots$$

$$\tilde{a}_{m1}x_1 + \tilde{a}_{m2}x_2 + \ldots + \tilde{a}_{mn}x_n \leq / = / \geq \tilde{b}_m$$

$$x_1, x_2, \ldots x_n \geq 0$$

8.4 METHODOLOGY

8.4.1 Crisp single-objective optimization model (CSOOM)

The general form of CSOOM with n number of variables and m number of constraints is written as follows:

$$Min \quad c_1 x_1 + c_2 x_2 + \ldots + c_n x_n$$

$$s.t. \quad a_{11} x_1 + a_{12} x_2 + \ldots + a_{1n} x_n \leq / = / \geq b_1$$

CSOOM-1: $\quad a_{21} x_1 + a_{22} x_2 + \ldots + a_{2n} x_n \leq / = / \geq b_2$

$$\cdots\cdots\cdots\cdots\cdots\cdots\cdots\cdots\cdots\cdots\cdots\cdots$$

$$a_{m1} x_1 + a_{m2} x_2 + \ldots + a_{mn} x_n \leq / = / \geq b_m$$

$$x_1, x_2, \ldots x_n \geq 0$$

In CSOOM-1, all the coefficients of objective functions and constraints are deterministic in nature. So, by using the Lingo software, we find the optimal solution for the above CSOOM-1.

8.4.2 Fuzzy single-objective optimization model (FSOOM)

A crisp optimization model can be fuzzy if the objective function, constraints, or both are fuzzy. The corresponding FSOOM can be written as follows:

$$Min \quad \tilde{c}_1 x_1 + \tilde{c}_2 x_2 + \ldots + \tilde{c}_n x_n$$

$$s.t. \quad \tilde{a}_{11} x_1 + \tilde{a}_{12} x_2 + \ldots + \tilde{a}_{1n} x_n \leq / = / \geq \tilde{b}_1$$

FSOOM-1: $\quad \tilde{a}_{21} x_1 + \tilde{a}_{22} x_2 + \ldots + \tilde{a}_{2n} x_n \leq / = / \geq \tilde{b}_2$

$$\cdots\cdots\cdots\cdots\cdots\cdots\cdots\cdots\cdots\cdots\cdots\cdots$$

$$\tilde{a}_{m1} x_1 + \tilde{a}_{m2} x_2 + \ldots + \tilde{a}_{mn} x_n \leq / = / \geq \tilde{b}_m$$

$$x_1, x_2, \ldots x_n \geq 0$$

Now to defuzzify the above FSOOM-1, we used the signed-distance method.

The above FSOOM-1 can be converted into a crisp model using the signed-distance method. Hence, the crisp model of FSOOM-1 can be given by

$$Min \quad d(c_1,0)x_1 + d(c_2,0)x_2 + \dots + d(c_n,0)x_n$$

$$s.t. \quad d(a_{11},0)x_1 + d(a_{12},0)x_2 + \dots$$
$$+ d(a_{1n},0)x_n \leq / = / \geq d(b_1,0)$$

CSOOM-2:
$$d(a_{21},0)x_1 + d(a_{22},0)x_2 + \dots$$
$$+ d(a_{2n},0)x_n \leq / = / \geq d(b_2,0)$$

$$\dots\dots\dots\dots\dots\dots\dots\dots\dots\dots\dots\dots\dots\dots$$

$$d(a_{m1},0)x_1 + d(a_{m2},0)x_2 + \dots$$
$$+ d(a_{mn},0)x_n \leq / = / \geq d(b_m,0)$$

$$x_1, x_2,\dots x_n \geq 0$$

Then by using Lingo software, we can obtain the optimal solution of the CSOOM-2.

8.4.3 Crisp multi-objective optimization model (CMOOM)

The general form of CMOOM with n number of variables and m number of constraints is written as follows:

$$Min \quad z_1(x), z_2(x), \dots, z_n(x)$$

$$s.t. \quad a_{11}x_1 + a_{12}x_2 + \dots + a_{1n}x_n \leq b_1$$

CMOOM-1:
$$a_{21}x_1 + a_{22}x_2 + \dots + a_{2n}x_n \leq b_2$$

$$\dots\dots\dots\dots\dots\dots\dots\dots\dots\dots\dots\dots\dots\dots$$

$$a_{m1}x_1 + a_{m2}x_2 + \dots + a_{mn}x_n \leq b_m$$

$$x_1, x_2 \geq 0$$

The above CMOOM can be converted into CSOOM using the weighted sum method (WSM).

8.4.4 Weighted sum method (WSM)

Only the minimizing problem can be done by using WSM. So, if we have given the maximizing problem first, we have to convert it into minimizing the problem.

The optimization problem is converted to a minimizing problem using the following relation.

Max Z = – Min (-Z)

As there are multiple objectives in the multi-objective optimization problem, we have to convert it into a single-objective optimization problem by using WSM as follows:

$$Min\ z \quad w_1 z_1 + w_2 z_2 + \ldots + w_n x_n$$

$$s.t. \quad a_{11} x_1 + a_{12} x_2 + \ldots + a_{1n} \leq / = / \geq b_1$$

$$a_{21} x_1 + a_{22} x_2 + \ldots + a_{2n} x_n \leq / = / \geq b_2$$

CSOOM-3: $\quad \ldots\ldots\ldots\ldots\ldots\ldots\ldots\ldots\ldots\ldots\ldots\ldots\ldots$

$$a_{m1} x_1 + a_{m2} x_2 + \ldots + a_{mn} x_n \leq / = / \geq b_m$$

$$w_1 + w_2 + \ldots + w_n = 1$$

$$0 \leq w_1, w_2, \ldots, w_n \leq 1$$

Then, the solution to the CSOOM-3 can be found by using Lingo software.

8.4.5 Fuzzy multi-objective optimization model (FMOOM)

Here, we consider the coefficients of objective function and constraints are fuzzy pentagonal numbers. The general form of FMOOM can be written as follows:

$$Min \quad \tilde{z}_1(x), \tilde{z}_2(x), \ldots \tilde{z}_n(x)$$

$$s.t. \quad \tilde{a}_{11} x_1 + \tilde{a}_{12} x_2 + \ldots + \tilde{a}_{1n} x_n \leq / = / \geq \tilde{b}_1$$

FMOOM-1: $\quad \tilde{a}_{21} x_1 + \tilde{a}_{22} x_2 + \ldots + \tilde{a}_{2n} x_n \leq / = / \geq \tilde{b}_2$

$$\ldots\ldots\ldots\ldots\ldots\ldots\ldots\ldots\ldots\ldots\ldots\ldots\ldots$$

$$\tilde{a}_{m1} x_1 + \tilde{a}_{m2} x_2 + \ldots + \tilde{a}_{mn} x_n \leq / = / \geq \tilde{b}_m$$

$$x_1, x_2, \ldots x_n \geq 0$$

In the FMOOM-1 model, first, we used the signed-distance method to defuzzify. Then, using the WSM method, the above model can be converted into a single-objective optimization model. And finally, the optimal solution can be obtained by using Lingo software.

8.5 NUMERICAL EXAMPLES

Case 1 (Example for CSOOM):

Here, we have considered a linear CSOOM with two variables and two constraints, and the coefficient of objective function and constraints are crisp in nature.

Let $c_1 = 6, c_2 = 4, a_{11} = 2, a_{12} = 1, a_{21} = 1, a_{22} = 1, b_1 = 10, b_2 = 8$.

Hence, the corresponding CSOOM model is given by:

$$Min\ z \quad 6x_1 + 4x_2$$

CSOOM-4:   *s.t.* $2x_1 + x_2 \geq 10$

$$x_1 + x_2 \geq 8$$

$$x_1, x_2 \geq 0$$

Sol^n: By using Lingo software, we get the optimal solution for the CSOOM-4

$x_1 = 2$
$x_2 = 6$
$z = 36$

Case 2 (Example for FSOOM):

Here, we have considered a linear FSOOM with two variables and two constraints, and the coefficient of objective function and constraints are fuzzy in nature. Here, we considered the values of $c_1, c_2, a_{11}, a_{12}, a_{21}, a_{22}, b_1, b_2$ as the fuzzy pentagonal number and are about 6, 4, 1, 1, 1, 10, and 8, respectively. Hence, the corresponding FSOOM model is given by:

$$Min\ z \quad (5.8, 5.8, 6, 6.1, 6.3)x_1 + (3.7, 3.9, 4, 4.1, 4.2)x_2$$

FSOOM-2:   *s.t.* $(1.7, 1.9, 2, 2.1, 2.2)x_1 + (0.8, 0.9, 1, 1.1, 1.2)x_2$

$$\geq (9.8, 9.9, 10, 10.1, 10.2)$$

$$(0.8, 0.9, 1, 1.1, 1.2)x_1 + (0.8, 0.9, 1, 1.1, 1.2)x_2$$
$$\geq (7.8, 7.9, 8, 8.1, 8.2)$$

$$x_1, x_2 \geq 0$$

Sol^n: Defuzzifying the objective function and constraint by using the signed-distance method, the deterministic equivalent of the FSOOM-2 is given by

Table 8.1 Optimal Solution Using Lingo Software

	CSOOM	FSOOM
x_1	2	2.024194
x_2	6	5.978506
z	36	36.01677

CSOOM-5:

$$Min\ z \quad 6.013x_1 + 3.988x_2$$
$$s.t. \quad 1.988x_1 + x_2 \geq 10.013$$
$$x_1 + x_2 \geq 8$$
$$x_1, x_2 \geq 0$$

We get the optimal solution to the above problem by using Lingo software (Table 8.1).

$x_1 = 2.024194$
$x_2 = 5.978506$
$z = 36.01677$

Case 3 (Example for CMOOM):

Here, we have considered a nonlinear CMOOM with two objective functions, two variables, and two constraints, and the coefficient of the objective function and constraints are crisp in nature.

CMOOM-1:

$$Min\ z_1 \quad 3x_1^2 + x_2^2$$
$$Min\ z_2 \quad \left(x_1^2 - 1\right)^2 + 3\left(x_2 - 1\right)^2$$
$$s.t. \quad -2 \leq x_1 \leq 2$$
$$-2 \leq x_2 \leq 2$$

Solution: By using the weighted sum method, we can convert the CMOOM-1 into CSOOM as follows:

CSOOM-6:

$$Min\ z \quad w_1\left(3x_1^2 + x_2^2\right) + w_2\left[\left(x_1^2 - 1\right)^2 + 3\left(x_2 - 1\right)^2\right]$$
$$s.t. \quad -2 \leq x_1 \leq 2$$
$$-2 \leq x_2 \leq 2$$
$$w_1 + w_2 = 1$$
$$w_1, w_2 \geq 0$$

Now we get the following optimal values using Lingo software (Table 8.2).

Table 8.2 Optimal Values Using Lingo Software

w_1	w_2	x_1	x_2	z_1	z_2	z
0.0	1.0	1.00	1.00	4.00	0.00	0.00
0.1	0.9	0.75	0.96	2.61	0.66	0.32
0.2	0.8	0.57	0.92	1.83	0.20	0.52
0.3	0.7	0.43	0.86	1.33	0.36	0.65
0.4	0.6	0.33	0.81	1.00	0.54	0.72
0.5	0.5	0.25	0.75	0.75	0.75	0.75
0.6	0.4	0.18	0.66	0.54	1.00	0.72
0.7	0.3	0.12	0.56	0.36	1.33	0.65
0.8	0.2	0.07	0.42	0.20	1.83	0.52
0.9	0.1	0.05	0.25	0.06	2.61	0.32
1.0	0.0	0.00	0.00	0.00	4.00	0.00

Case 4 (Example for FMOOM):

Here, we have considered a nonlinear FSOOM with two objective functions, two variables, and two constraints, and the coefficient of the objective function and constraints are fuzzy in nature. And the coefficient of the objective function and coefficient of constraint are taken as fuzzy pentagonal numbers. Now the corresponding FMOOM is given as follows:

$$Min\ z_1 \quad (2.8,2.9,3,3.2,3.3)\,x_1^2 + (0.9,1,1.1,1.2,1.3)\,x_2^2$$

$$Min\ z_2 \quad \big((0.9,1,1.1,1.2,1.3)\,x_1^2 - (0.9,1,1.1,1.2,1.3)\big)^2$$

$$+ (2.8,2.9,3,3.2,3.3)\big((0.9,1,1.1,1.2,1.3)\,x_2$$

$$- (0.9,1,1.1,1.2,1.3)\big)^2$$

FMOOM-2:   *s.t.* $-(1.7,1.9,2,2.1,2.2) \le x_1 \le (1.7,1.9,2,2.1,2.2)$

$$-(1.7,1.9,2,2.1,2.2) \le x_2 \le (1.7,1.9,2,2.1,2.2)$$

$$x_1, x_2 \ge 0$$

To solve this problem, we first have to convert this fuzzy model into a crisp one using the signed-distance method for the pentagonal fuzzy number.

Hence, the deterministic equivalent of the above fuzzy model can be written as:

$$Min\ z_1 \quad 3.038x_1^2 + 1.1x_2^2$$

$$Min\ z_2 \quad \left(1.1x_1^2 - 1.1\right)^2 + 3.038\left(1.1x_2 - 1.1\right)^2$$

CMOOM-2:   *s.t.* $-1.988 \le x_1 \le 1.988$

$$-1.988 \le x_2 \le 1.988$$

$$x_1, x_2 \ge 0$$

Using the weighted sum method

$$Min\ z \quad w_1\left(3.038x_1^2 + 1.1x_2^2\right)$$

$$+ w_2\left[\left(1.1x_1^2 - 1.1\right)^2 + 3.038\left(1.1x_2 - 1.1\right)^2\right]$$

CSOOM-7:   *s.t.* $-1.988 \le x_1 \le 1.988$

$$-1.988 \le x_2 \le 1.988$$

$$w_1 + w_2 = 1$$

$$0 \le w_1, w_2 \le 1$$

Now we get the following optimal values using Lingo software (Table 8.3).

8.6 CONCLUSION

We solved the optimization problem with the aforementioned analysis for single-objective and multi-objective models. The models are developed in both crisp and fuzzy environments. In a fuzzy model, all parameters are assumed to be fuzzy pentagonal numbers. We used signed-distance method for defuzzification. The weighted sum method converts the multi-objective

Table 8.3 Optimal Values Using Lingo Software

w_1	w_2	x_1	x_2	z_1	z_2	z
0.0	1.0	1.00	0.00	4.00	0.00	0.00
0.1	0.9	0.71	0.96	2.59	0.09	0.30
0.2	0.8	0.56	0.92	1.91	0.24	0.51
0.3	0.7	0.44	0.88	1.45	0.42	0.66
0.4	0.6	0.34	0.83	1.11	0.62	0.74
0.5	0.5	0.26	0.75	0.85	0.85	0.78
0.6	0.4	0.19	0.68	0.63	1.14	0.77
0.7	0.3	0.13	0.58	0.43	1.53	0.71
0.8	0.2	0.08	0.45	0.24	2.1	0.58
0.9	0.1	0.03	0.26	0.08	3.06	0.36
1.0	0.0	0.00	0.00	0.00	4.00	0.00

optimization model into a single-objective optimization problem. Finally, we find the optimal solution for the crisp model by using Lingo software.

REFERENCES

[1] Zadeh, L.A. [1965]. Fuzzy sets, *Information Control*, 8.

[2] Zimmermann, H.J. [1996]. *Fuzzy Set Theory and Its Application*, 3rd ed., Kluwer, Academic Publishers.

[3] Yao, Jing-Shing, Kweimei Wu [2000]. Ranking fuzzy numbers based on decomposition principle and signed distance, *Fuzzy Sets, and Systems*, 116 (2), 275–288.

[4] Yao, J.S., S.C. Chang, J.S. Su [2003]. Fuzzy inventory without backorder with fuzzy sense for fuzzy total cost and fuzzy storing cost defuzzified by centroid and signed distance, *European Journal of Operational Research*, 148, 401–409.

[5] Syed, J.K., L.A. Aziz [2007]. Fuzzy inventory model without shortages using signed distance method, *Applied Mathematics and Information Sciences*, 1 (2), 203–209.

[6] Chen, S.H., S.M. Chang [2008]. Optimization of fuzzy production inventory model with unrepairable defective products, *International Journal of Production Economics*, 113 (2), 887–894.

[7] Miettinen, Kaisa [1973]. *Multi-Objective Optimization, Interactive and Evolutionary Approaches*, University of Jyväskylä, Department of Mathematics Information Technology.

[8] Das, I., J.E. Dennis [1997]. A closer look at drawbacks of minimizing weighted sums of objectives for Pareto set generation in multicriteria optimization problem, *Structural and Multidisciplinary Optimization*, 14, 63–69.

[9] R. Timothy Marler, Jasbir S. Arora [2010]. The weighted sum method for multi-objective optimization: New insights, *Structural and Multidisciplinary Optimization*, 41, 853–862.

[10] Ramli, S., S.H. Jaaman [2018]. Optimal solution of fuzzy optimization using pentagonal fuzzy number, *AIP Conference Proceedings*, 1974 (1). doi:10.1063/1.5041597.

[11] Das, S.K., A. Chakrabort [2021]. A new approach to evaluate linear programming problem in pentagonal neutrosophic environment, *Complex & Intelligent Systems*, 7, 101–110.

[12] Alharbi, M.G., H.W. Khalifa [2021]. On a flow shop scheduling problem with fuzzy pentagonal processing time, *Journal of Mathematics*, 2021, 7.

Chapter 9

Sensitivity analysis of fuzzy linear programming

Sanjaya K. Behera and Jyotiranjan Nayak

9.1 INTRODUCTION

Linear Programming (LP) is one of the most frequently applied techniques in modeling the real-world problems. Traditional LP requires deterministic and precise data for finding optimal solution. To deal with vague and imprecise data, we often use different fuzzy domains. There exist efficient fuzzy models which approximate the vagueness in real life systems presented by Zadeh [1]. Based on the theory of fuzzy sets, fuzzy decision making [2, 3] and fuzzy mathematical programming have been widely used to tackle problems encountered in many different real-world applications, that is, agricultural economics, network locations, banking and finance, environment management, inventory management, manufacturing and productions, resource allocation, supply chain management, transportation management, product mix, and marketing.

The concepts of decision making in fuzzy environment are proposed by Zadeh [1] for further improvisation. Tanaka et al. [4] have considered the linear programming with fuzzy constraints. Zimmerman [2] has given an algorithm to solve multi-objective FLP models. Extending that algorithm, many authors have solved the FLP models in different ways. Ebrahimnejad [5] has solved the FLP models by using the simplex method. Ebrahimnejad et al. [6] have solved the FLP models with symmetric trapezoidal numbers by using the simplex method with the help of ranking numbers.

Some algorithms are developed to solve FLP models to determine fuzzy compromise solutions by Behera et al. [7]. After getting the solution of the FLP models, sensitivity analysis of the solution has been studied in fuzzy domain. Chel et al. [8] and Kala [9] have considered the variables under sensitivity analysis to vary within a real interval for the better solution in the model. We are assuming the variables to vary within the fuzzy intervals.

This chapter is organized as follows. We review some necessary concepts and background on fuzzy arithmetic in Section 9.2. In Section 9.3, we have discussed the variation of requirement vector and cost coefficient in FLP models. In Section 9.4, we have presented a numerical example to study the variation of the variables. Finally, the conclusion and future work are mentioned in Section 9.5.

DOI: 10.1201/9781003462422-9

9.2 PRELIMINARIES

In this section, we review some necessary concepts and backgrounds on fuzzy arithmetic:

Definition 2.1: A fuzzy set $\tilde{A}$ on R is called a symmetric trapezoidal fuzzy number if its membership function is defined as follows: Kaufman et al. [10]:

$$\mu_{\tilde{A}}(x) = \begin{cases} \dfrac{x-(a^L-\alpha)}{\alpha} & if \ \ a^L-\alpha \leq x \leq a^L \\ 1 & if \ \ a^L \leq x \leq a^U \\ \dfrac{(a^U+\alpha)-x}{\alpha} & if \ \ a^U \leq x \leq a^U+\alpha \\ 0 & else \end{cases}$$

We denote a symmetric trapezoidal fuzzy number $\tilde{A}$ by $\tilde{A} = \left(a^L, a^U, \alpha, \alpha\right)$ and the set of all symmetric trapezoidal fuzzy numbers by F(R).

Definition 2.2: The arithmetic operations on two symmetric trapezoidal fuzzy numbers

$\tilde{A} = \left(a^L, a^U, \alpha, \alpha\right)$ and $\tilde{B} = \left(b^L, b^U, \beta, \beta\right)$ are given by [11]:

$\tilde{A} + \tilde{B} = \left(a^L + b^L, a^U + b^U, \alpha + \beta, \alpha + \beta\right),$

$\tilde{A} - \tilde{B} = \left(a^L - b^U, a^U - b^L, \alpha + \beta, \alpha + \beta\right)$

$$A\tilde{B} = \left(\left(\frac{a^L+b^L}{2}\right)\left(\frac{a^U+b^U}{2}\right) - t, \right.$$
$$\left(\frac{a^L+b^L}{2}\right)\left(\frac{a^U+b^U}{2}\right) + t,$$
$$\left. \left|a^U\beta + b^U\alpha\right|, \ \left|a^U\beta + b^U\alpha\right| \right)$$

$where \ t = \dfrac{t_2-t_1}{2},$

$t_1 = \min\left\{a^L b^L, a^L b^U, a^U b^L, a^U b^U\right\},$

$t_2 = \max\left\{a^L b^L, a^L b^U, a^U b^L, a^U b^U\right\}$

$k\tilde{A} = \begin{cases} (ka^L, ka^U, k\alpha, k\alpha) & if \ k \geq 0 \\ (ka^U, ka^L, -k\alpha, -k\alpha) & if \ k < 0 \end{cases}$

Definition 2.3: Let $\tilde{A} = \left(a^L, a^U, \alpha, \alpha\right)$ and $\tilde{B} = \left(b^L, b^U, \beta, \beta\right)$ be two symmetric trapezoidal fuzzy numbers [10]. The relations $\lesssim$ and $\approx$ are defined as follows:

$\tilde{A} \lesssim \tilde{B}$ if and only if:

(i) $\dfrac{(a^L - \alpha) + (a^U - \alpha)}{2} < \dfrac{(b^L - \beta) + (b^U - \beta)}{2}$, *that is* $\dfrac{a^L + a^U}{2} < \dfrac{b^L + b^U}{2}$

(*in this case we may write* $\tilde{A} \lesssim \tilde{B}$), *or*

(ii) $\dfrac{a^L + a^U}{2} = \dfrac{b^L + b^U}{2}, b^L < a^L, a^U < b^U$ (*in this case we say* $\tilde{A} \approx \tilde{B}$), *or*

(iii) $\dfrac{a^L + a^U}{2} = \dfrac{b^L + b^U}{2}, b^L = a^L, a^U = b^U$ (*in this case we say* $\tilde{A} \approx \tilde{B}$).

Definition 2.4: Ranking Function, Kaino et al. [11]: The ranking function is defined on set of real numbers **R**, which maps each fuzzy number into the real line, where a natural order exists, that is, $R: F(\mathbf{R}) \to \mathbf{R}$, where $F(\mathbf{R})$ is a set of fuzzy numbers. Let $\tilde{A} = (a^L, a^U, \alpha, \alpha)$ be a symmetric fuzzy trapezoidal fuzzy number. A standard the ranking function $R(\tilde{A}) = \dfrac{a^L + a^U}{2}$ has been used by Ebrahimnejad et al. [12]. The same function is used by us in the results. For more in-depth study, readers can refer the research outcomes outlined in research papers [13–20].

Note: $\tilde{A} \lesssim \tilde{B}$ *iff* $R(\tilde{A}) \lesssim R(\tilde{B})$

Definition 2.5: Linear programming:
A linear programming (LP) problem is defined as:

Maximize $z = cx$
subject to $Ax = b$
$$x \geq 0$$
where $b = (b_1, b_2, ..., b_m)^T, c = (c_1, c_2, ..., c_n)$ *and* $A = [a_{ij}]_{m \times n}$.

Definition 2.6: Fuzzy Linear programming:
Suppose that in the linear programming problem the parameters, that is, coefficients of objective functions, elements of coefficient matrix, elements of requirement vectors, and unknown variable are fuzzy number. Then the linear programming problems is fuzzy linear programming problem.

Definition 2.7: Fuzzy feasible solution:
The vector $x \in R_n$ is a feasible solution to FLP if and only if x satisfies the constraints of the problem.

Definition 2.8: Fuzzy optimal solution:
A feasible solution x^* is an optimal solution for FLP, if for all feasible solution x for FLP, then $\tilde{c}x^* \gtrsim \tilde{c}x$.

9.3 EXAMPLES

Example 3.1

Maximize $\tilde{Z} = (13,15,2,2)\tilde{x}_1 + (12,14,3,3)\tilde{x}_2 + (15,17,2,2)\tilde{x}_3$

subject to

$$12\tilde{x}_1 + 13\tilde{x}_2 + 12\tilde{x}_3 \tilde{\leq} (475,505,6,6)$$
$$14\tilde{x}_1 + 13\tilde{x}_3 \tilde{\leq} (460,480,8,8)$$
$$12\tilde{x}_1 + 15\tilde{x}_2 \tilde{\leq} (465,495,5,5)$$
$$\tilde{x}_1, \tilde{x}_2, \tilde{x}_3 \tilde{\geq} \tilde{0}$$

Solution: Using simplex method, we solve the given FLP model.

Since the ranking value of all the variables in z-row is positive, therefore, an optimal solution has been obtained. The optimal solution is

$$\tilde{x}_2 = \left(\frac{405}{169}, \frac{1045}{169}, \frac{86}{169}, \frac{86}{169}\right), \quad \tilde{x}_3 = \left(\frac{460}{13}, \frac{480}{13}, \frac{8}{13}, \frac{8}{13}\right),$$

$$\tilde{x}_6 = \left(\frac{62910}{169}, \frac{77430}{169}, \frac{3455}{169}, \frac{3455}{169}\right)$$

and the maximum of $\tilde{Z} = \left(\frac{62910}{169}, \frac{77430}{169}, \frac{3455}{169}, \frac{3455}{169}\right)$.

9.4 SENSITIVITY ANALYSIS

9.4.1 Variation in the requirement vector

Let $\tilde{X}_B$ be the optimum basic feasible solution of the FLPP:

Maximize $\tilde{z} \cong \tilde{C}\tilde{X}$

subject to the constraints

$A\tilde{X} \cong \tilde{b}$ and $\tilde{x} \tilde{\geq} \delta$, where A is a $m \times n$ real matrix.

Table 9.1 Steps of the simplex method in solving FLP model I

B	x_1	x_2	x_3	x_4	x_5	x_6	X_B	$R(X_B)$	Ratio
$\tilde{Z}$	-(13,15, 2,2)	-(12,14, 3,3)	-(15,17, 2,2)	$\tilde{0}$	$\tilde{0}$	$\tilde{0}$	$\tilde{0}$		
$R(\tilde{Z})$	-14	-13	-16	0	0	0	0	0	
$\tilde{x}_4$	12	13	12	1	0	0	(475,505,6,6)	490	-
$\tilde{x}_5$	14	0	13	0	1	0	(460,480,8,8)	470	470/12
$\tilde{x}_6$	12	15	0	0	0	1	(465,495,5,5)	480	480/13

Table 9.2 Steps of the simplex method in solving FLP model 2

B	x_1	x_2	x_3	x_4	x_5	x_6	X_B	$R(X_B)$	Ratio
$\tilde{Z}$	$\left(\dfrac{15}{13},\dfrac{69}{13},\dfrac{54}{13},\dfrac{54}{13}\right)$	$(-14,-12,3,3)$	$\tilde{0}$	$\tilde{0}$	$\left(\dfrac{15}{13},\dfrac{17}{13},\dfrac{2}{13},\dfrac{2}{13}\right)$	$\tilde{0}$	$\left(\dfrac{6890}{13},\dfrac{8150}{13},\dfrac{1096}{13},\dfrac{1096}{13}\right)$		
$R(\tilde{Z})$									
$\tilde{x}_4$	$-12/13$	13	0	1	$-12/13$	0	$\left(\dfrac{415}{13},\dfrac{1045}{13},\dfrac{86}{13},\dfrac{86}{13}\right)$	$\dfrac{1460}{26}$	$-$
$\tilde{x}_3$	$14/13$	0	1	0	$1/13$	0	$\left(\dfrac{460}{13},\dfrac{480}{13},\dfrac{8}{13},\dfrac{8}{13}\right)$	$\dfrac{1940}{26}$	4.319
$\tilde{x}_6$	12	15	0	0	0	1	$(465,495,5,5)$	480	32

Table 9.3 Steps of the simplex method in solving FLP model 3

B	x_1	x_2	x_3	x_4	x_5	x_6	X_B
$\tilde{Z}$	$\left(\dfrac{51}{169},\dfrac{729}{169},\dfrac{738}{169},\dfrac{738}{169}\right)$	$\tilde{0}$	$\tilde{0}$	$\left(\dfrac{14}{13},\dfrac{12}{13},\dfrac{3}{13},\dfrac{3}{13}\right)$	$\left(\dfrac{51}{169},\dfrac{53}{169},\dfrac{62}{169},\dfrac{62}{169}\right)$	$\tilde{0}$	$\left(\dfrac{6890}{13},\dfrac{8150}{13},\dfrac{1096}{13},\dfrac{1096}{13}\right)$
$R(\tilde{Z})$							
$\tilde{x}_2$	$-12/169$	1	0	$1/13$	$-12/169$	0	$\left(\dfrac{405}{169},\dfrac{1045}{169},\dfrac{86}{169},\dfrac{86}{169}\right)$
$\tilde{x}_3$	$14/13$	0	1	0	$1/13$	0	$\left(\dfrac{460}{13},\dfrac{480}{13},\dfrac{8}{13},\dfrac{8}{13}\right)$
$\tilde{x}_6$	$1848/169$	0	0	$-15/13$	$180/169$	1	$\left(\dfrac{62910}{169},\dfrac{77430}{169},\dfrac{3455}{169},\dfrac{3455}{169}\right)$

Any change in the requirement vector $\tilde{b}$ does not affect the optimality conditions.

However, it may affect the feasibility conditions. Thus, if $\tilde{b}_k$ is changed to $\tilde{b}_k + \Delta\tilde{b}_k$, the new basic feasible solution is given by

$$\tilde{X}_B^* \cong \tilde{B}^{-1}\tilde{b}^* \text{ where } \tilde{b}^* \cong [\tilde{b}_1, \tilde{b}_2 \cdots \tilde{b}_k + \Delta\tilde{b}_k, \tilde{b}_n]$$

i.e., $\tilde{X}_B^* \cong \tilde{B}^{-1}\tilde{b} + \tilde{B}^{-1}[0,0,\cdots,\Delta\tilde{b}_k,0]$

$$\cong \tilde{X}_B + \vec{\tilde{B}}_k\Delta\tilde{b}_k \text{ where } \vec{\tilde{B}}_k \text{ is called the } k^{th} \text{ column vector of } \tilde{B}^{-1}.$$

In order to maintain the feasibility of the solution at each iteration, $\tilde{X}_B^*$ must be fuzzy non negative.

Thus,

$$\tilde{X}_{B_i} + B_{ik}\Delta\tilde{b}_k \tilde{\ge} 0$$

$$\Delta\tilde{b}_k \tilde{\ge} -\frac{\tilde{X}_{B_i}}{B_{ik}} \text{ for } B_{ik} \tilde{\ge} 0$$

$$\Delta\tilde{b}_k \tilde{\le} -\frac{\tilde{X}_{B_i}}{B_{ik}} \text{ for } B_{ik} \tilde{\le} 0.$$

9.4.2 Variation in the cost vector

Let $\tilde{C}_B$ be the optimum basic feasible solution of the FLPP:

$$\text{Maximize } \tilde{z} \cong \tilde{C}\tilde{X}$$

subject to

$$A\tilde{X} \cong \tilde{b} \text{ and } \tilde{x} \tilde{\ge} 0.$$

Here A is any $m \times n$ nonzero real matrix.

Let $\Delta\tilde{C}_k$ be the amount which is added to the kth component $\tilde{C}_k$ of $\tilde{C} \cong (\tilde{c}_1, \cdots \tilde{c}_k, \cdots \tilde{c}_n)$.

The new value of the kth component becomes $\tilde{C}_k^* \cong \tilde{C}_k + \Delta\tilde{C}_k$.

Since $\tilde{X}_B \cong B^{-1}\tilde{b}$ is independent of $\tilde{C}$, any change in the same component $\tilde{C}_j$, $j = 1,2,\cdots,n$ of $\tilde{C}$ will not affect the value of $\tilde{X}_B$ and hence the current solution $\tilde{X}_B$ will always remain basic feasible.

Now if $\tilde{C}_B$ be the cost vector associated with the optimum basic feasible solution, then any change in $\tilde{C}$ may definitely bring some change in the optimality conditions, because

$$\tilde{Z}_k - \tilde{C}_k \cong \tilde{C}_B y_k \ (y_k \text{ is the dual variable})$$

Now there are two possibilities;

(i) $\tilde{C}_k$ is not in $\tilde{C}_B$

(ii) $\tilde{C}_k$ is in $\tilde{C}_B$

Case (i): Here $\tilde{C}_k$ is not in $\tilde{C}_B$, that is, $\tilde{C}_k$ is not the coefficient of basic variable in objective. The net evaluation corresponding to the non-basic variable $\tilde{X}_k$ is given by $\tilde{Z}_k - \tilde{C}_k \cong \tilde{Z}_k - (\tilde{c}_k + \Delta\tilde{c}_k)$ for all $\tilde{c}_k \notin \tilde{C}_B$.

Since $\tilde{Z}_k$ is not affected for any change in $\tilde{c}_k$, therefore, the current solution $\tilde{X}_B$ remains optimum for the new problem if $\tilde{Z}_k - (\tilde{c}_k + \Delta\tilde{c}_k) \gtrsim 0$ or $\Delta\tilde{C}_k \lesssim \tilde{Z}_k - \tilde{C}_k$.

Since $\tilde{z} \cong \tilde{C}_B \tilde{X}_B$ is independent of $\tilde{C}_k$, the value of the objective function and fuzzy optimum solution will remain unchanged.

Case (ii): Here $\tilde{C}_k$ is in $\tilde{C}_B$. The net evaluation corresponding to the basic variable $\tilde{X}_k$ is given by

$$\tilde{Z}_k^* - \tilde{C}_k \cong \tilde{C}_B^* y_k - \tilde{C}_j$$

$$\tilde{Z}_k^* - \tilde{C}_k \cong \sum_{\substack{i-1 \\ i \neq k}}^{m} \tilde{C}_{B_i} y_{ij} + \left(\tilde{C}_{BK} + \Delta\tilde{C}_{BK} \right) y_{kj} - \tilde{C}_j$$

$$\cong \sum_{i-1}^{m} \tilde{C}_{B_i} y_{ij} - \tilde{C}_j + \Delta\tilde{C}_{BK} y_{kj}$$

$$\cong \left(\tilde{Z}_j - \tilde{C}_j \right) + \Delta\tilde{C}_{BK} y_{kj}.$$

For the current basic feasible solution to remain optimum, we must have

$$\tilde{Z}_k^* - \tilde{C}_j \gtrsim \tilde{0}$$

$$\Rightarrow \left(\tilde{Z}_k^* - \tilde{C}_j \right) + \Delta\tilde{C}_K y_{kj} \gtrsim \tilde{0}$$

$$\Rightarrow \Delta\tilde{C}_K \gtrsim - \left\{ \frac{\tilde{Z}_j - \tilde{C}_j}{y_{kj}} \right\} \ if \ y_{kj} \geq 0,$$

$$\text{and } \Delta\tilde{C}_K \lesssim \frac{-\tilde{Z}_j - \tilde{C}_j}{y_{kj}} if \ y_{kj} \leq 0.$$

So, $Max -\left\{\dfrac{\tilde{Z}_j -\tilde{C}_j}{y_{kj} > 0}\right\} \precsim \Delta\tilde{C}_k \precsim Min\left\{\dfrac{-(\tilde{Z}_j -\tilde{C}_j)}{y_{kj} < 0}\right\}.$

Thus, the revised value of the objective function is

$$Z^* \cong \sum_{\substack{i=1\\i\neq k}}^{m} \tilde{C}_{B_i}\tilde{x}_{B_i} + \left(\tilde{C}_{BK} + \Delta\tilde{C}_{BK}\right)$$

$$\cong \tilde{Z} + \tilde{x}_k\Delta\tilde{c}_k$$

Variation in $\tilde{C}_1$:

Since $\tilde{C}_1 \notin \tilde{C}_B$, the change $\Delta\tilde{C}_1$ *in* $\tilde{C}_1$ so that solution remains optimal is given by

$$\Delta\tilde{C}_1 \precsim \tilde{Z}_1 -\tilde{C}_1,$$

$$\Rightarrow \Delta\tilde{C}_1 \precsim \left(\frac{51}{169},\frac{729}{169},\frac{738}{169},\frac{738}{169}\right).$$

Change in $\tilde{C}_1$ such that the optimum solution remains unchanged is,

$$-\infty \precsim \tilde{C}_1 \precsim (13,15,2,2)+\left(\frac{51}{169},\frac{729}{169},\frac{738}{169},\frac{738}{169}\right).$$

Variation in $\tilde{C}_2$:

Since $\tilde{C}_2 \in \tilde{C}_B$, then change $\Delta\tilde{C}_2$ in $\tilde{C}_2$ so that solution remaining optimum is given by,

$$Max -\left\{\dfrac{\tilde{Z}_j -\tilde{C}_j}{y_{2j} > 0}\right\} \precsim \Delta\tilde{C}_2 \precsim Min\left\{\dfrac{-(\tilde{Z}_j -\tilde{C}_j)}{y_{2j} < 0}\right\}$$

$$Max -\left\{\dfrac{(14/13,12/13,3/13,3/13)}{1/13}\right\} \precsim \Delta\tilde{C}_2$$

$$\precsim Min -\left\{\dfrac{(51/169,729/169,738/169,738/169)}{-12/169},\right.$$

$$\left.\dfrac{(51/169,53/169,62/169,62/169)}{-12/169}\right\}$$

$$\Rightarrow \left(-12,-13,3,3\right) \tilde{\leq} \Delta \tilde{C}_2 \tilde{\leq} \left(\frac{51}{12},\frac{53}{12},\frac{62}{12},\frac{62}{12}\right).$$

Variation in $\tilde{C}_3$:

Since $\tilde{C}_3 \in \tilde{C}_B$, then change $\Delta \tilde{C}_3$ in $\tilde{C}_3$ so that solution remaining optimum is given by,

$$Max - \left\{ \frac{\tilde{Z}_j - \tilde{C}_j}{y_{3j} > 0} \right\} \tilde{\leq} \Delta \tilde{C}_3 \tilde{\leq} Min \left\{ \frac{-(\tilde{Z}_j - \tilde{C}_j)}{y_{3j} < 0} \right\}$$

$$Max - \left\{ \frac{\left(51/169,729/169,738/169,738/169\right)}{14/13}, \right.$$

$$\left. \frac{\left(51/169,53/169,\,62/169,62/169\right)}{1/13} \right\} \tilde{\leq} \Delta \tilde{C}_3 \tilde{\leq} \infty$$

$$\Rightarrow \left(-53/13,-51/13,\,62/13,\,62/13\right) \tilde{\leq} \Delta \tilde{C}_3 \tilde{\leq} \infty .$$

Variation in $\tilde{b}_1$:

$$\Delta \tilde{b}_k \tilde{\geq} - \frac{\tilde{X}_{B_i}}{B_{ik}} \; for \;\; B_{ik} \tilde{\geq} 0$$

$$\Delta \tilde{b}_k \tilde{\leq} - \frac{\tilde{X}_{B_i}}{B_{ik}} \; for \;\; B_{ik} \tilde{\leq} 0$$

$$\Delta \tilde{b}_1 \tilde{\leq} - \left(\frac{405/169,\,1045/169,\,86/169,\,86/169}{-12/169} \right) \; and$$

$$\Delta \tilde{b}_1 \tilde{\geq} - \left(\frac{405/169,\,1045/169,\,86/169,\,86/169}{1/169} \right).$$

9.5 CONCLUSION

In this chapter, we have used the simplex method to solve the FLP models with symmetric trapezoidal fuzzy numbers and then the condition on various cases of sensitivity analysis are discussed. The variables are assumed to lie within fuzzy intervals. This method is useful to get the compromise solution of different FLPs.

REFERENCES

[1] Zadeh L. A. (1965), Fuzzy sets, *Information and Control*, 8, 338–353.

[2] Zimmermann H. J. (1983), Using fuzzy sets in operational research, *European Journal of Operation Research*, 13, 201–216.

[3] Zimmermann H. J. (1987), *Fuzzy Sets, Decision Making and Expert Systems*, Kluwer Academic Publishers, Aston.

[4] Tanaka H., Asai K. (2000), Fuzzy linear programming problems with fuzzy numbers, *Fuzzy Sets and Systems*, 109, 21–33.

[5] Ebrahimnejad A. (2011), Sensitivity analysis in fuzzy number linear programming problems, *Mathematical and Computational Model*, 53, 1878–1888.

[6] Ebrahimnejad A., Tavana M. (2014), A novel method for solving linear programming problems with symmetric trapezoidal fuzzy numbers, *Applied Mathematical Modelling*, 38, 4388–4395.

[7] Behera S. K., Nayak J. R. (2011), Solution of multi-objective mathematical programming problems in fuzzy approach, *International Journal of Computer Science and Engineering*, 3(12), 3790–3799.

[8] Chen Y., Yu J., Khan S. (2010), Spatial sensitivity analysis of multi-criteria eights in GIS based land suitability evaluation, *Journal of Environment Modelling & Software*, 25, 1582–1591.

[9] Kala Z. (2008), *Sensitivity Analysis of Carrying Capacity of Steel Plane Frames to Imperfections*, Numerical Analysis and Applied Mathematics, AIP Conference Proceeding 1048, American Institute of Physics, Melville, 298–301.

[10] Kaufmann A., Gupta M. M. (1985), *Introduction to Fuzzy Arithmetic Theory and Applications*, Van Nostrand, Reinhold, New York.

[11] Kaino T., Hirota K. (1999), Y-CG derivative of fuzzy relation and its application to sensitivity analysis, *The International Journal of Fuzzy Systems*, 1(2), 129–132.

[12] Mahdavi-Amiri N., Nasseri S. H., Yazdani A. (2009), Fuzzy primal simplex algorithms for solving fuzzy linear programming problems, *Iranian Journal of Operational Research*, 1, 68–84.

[13] Behera S. K., Nayak J. R. (2012), Optimal solution of fuzzy non-linear programming problem with linear constraints, *International Journal of Advances in Science and Technology*, 4(2), 43–52.

[14] Bellman R. E., Zadeh L. A. (1970), Decision making in fuzzy environment, *Management Science*, 17, 141–164.

[15] Borogonovo E., Plischke E. (2016), Sensitivity analysis: A review of recent advances, *European Journal of Operation Research*, 248, 869–887.

[16] Marichal J. L. (2000), On Sugeno integral as an aggregation function, *Fuzzy Sets and Systems*, 114, 347–365.

[17] Nayak J. R. (2004), *Some problems of non-convex programming problems and the properties of some non-convex functions*, Ph. D Thesis, Utkal University, Odisha.

[18] Pattnaik M. (2013), Fuzzy multi-objective linear programming problems: Sensitivity analysis, *Journal of Mathematics and Computer Science*, 7, 131–137.

[19] Sugeno M. (1972), Fuzzy measure and fuzzy integral, *Journal of the Society of Instrument and Control Engineers*, 8(2), 218–226.

[20] Velasquez M., Hester P. T. (2013), An analysis of multi-criteria decision making methods, *International Journal of Operations Research*, 10, 56–66.

Chapter 10

Constrained effective bit selection algorithm for solving 0–1 knapsack problem

Rajpal Rajbhar and L. N. Das

10.1 INTRODUCTION

Martello [1] described a comprehensive review of the technique commonly used in solving the 0–1 Knapsack problem. There are three solution processes for the 0–1 knapsack problem: the exact solution method, the meta-heuristic method, and the heuristic method with the approximate solution. Concerning the exact method to solve the 0–1 knapsack problem, the dynamic programming algorithm, and branch and bound algorithm are the specifications used in these methods. The branch and bound algorithm were introduced in the year 1967 by Koelesar [2]. Later E. Horowitz [3], Pisinger [4], and Martello [5] expanded the branch and bound method algorithm to solve the 0–1 Knapsack problem concern to dynamic programming; the algorithm is P. Toth [6] was developed by A. Rong [7, 8].

A heuristic technique is applied in solving the optimization problem, which is found in the work of P. Vasant et al. [9]. A review of the literature survey concerning to 0–1 Knapsack problem-solving process also used the heuristic technique, whose detailed descriptions are found in the introduction of the paper of Jianhui et al. [10].

10.2 DESCRIPTION OF KNAPSACK 0–1

The Knapsack 0–1 can be formally defined as follows. Given an instance of the knapsack problem with an item set consisting of n items, the index j item has weight w_j and profit p_j, within the Knapsack occupying capacity value c. The objective is to select a subset of items among the n items, such that the profit of the selected item is maximized and the total weight does not exceed c. Mathematically, the problem is formulated as follows:

$$\text{Maximize } z = \sum_{j=1}^{n} p_j x_j \tag{10.1}$$

subject to $\sum_{j=1}^{n} w_j x_j \leq c$

$x_j \in \{0,1\}\ \ j = 1, 2 ... n$

We denote the optimal vector by $x^* = (x1^*, x2^* \text{-----} xn^*)$ and the optimal solution of z is z^*, the set X^* denoted the optimal solution set of items corresponding to the optimal solution vector.

10.2.1 Definition of greedy degree

A profitable packing of items into the Knapsack is an intuitive approach, and it would be to consider the profit-to-weight ratio e_j of each item which is also called the efficiency of the item with $e_j = p_j / w_j$ items generate the highest profit while consuming the lowest capacity, sorted by their efficiency in decreasing order. We apply the greedy algorithm at the start of item specifications and obtain a solution: the optimal local values rather than the optimal. Some items are included in the Knapsack by the greedy algorithm application. In other ways, the selective items are determinate, and unselected items are uncertain. Based on the consideration, a certain item in the Knapsack is a determinant item if it should be included in advance and never taken out from the Knapsack. Here the question is, which and how many items are determinants in the proposed model for the concept of the greedy degree? The answer to this question is found after applying the algorithm to decide the greedy degree. We define the greedy degree with the following expressions.

Definition: From the n items, the algorithm arranges in the greedy degree of m items if the m items are included already in the Knapsack by the satisfaction of the inequalities mentioned in the following manner. The relation parameters are described and mentioned in references [1, 11, 12], which is reviewed as follows:

$$\left\{ \frac{\sum_{j=1}^{k} w_j}{W^G} \wedge \frac{\sum_{j=1}^{k} p_j}{z^G} \leq \eta \right. \tag{10.2}$$

$$\left\{ \frac{\sum_{j=1}^{k} w_j}{W^G} \wedge \frac{\sum_{j=1}^{k} p_j}{z^G} > \eta \right. \tag{10.3}$$

$$\left\{ \frac{\sum_{j=1}^{n/2} w_j + \sum_{j=\frac{n}{2}+1}^{k} w_j}{W^G} \wedge \frac{\sum_{j=1}^{\frac{n}{2}} p_j + \sum_{j=\frac{n}{2}+1}^{k} p_j}{z^G} < \zeta \right. \tag{10.4}$$

$$\left\{ \frac{\sum_{j=1}^{n/2} w_j + \sum_{j=\frac{n}{2}+1}^{k} w_j}{W^G} \wedge \frac{\sum_{j=1}^{\frac{n}{2}} p_j + \sum_{j=\frac{n}{2}+1}^{k} p_j}{z^G} \geq \zeta \right. \tag{10.5}$$

Where Q, W^G, and z^G are the number of items obtained by the greedy algorithm, total weight of items, and objective function value, respectively. The parameters $\eta, \zeta \in (0,1)$ are used in the inequality to scale the serial number of items. The greedy degree deciding algorithm is designed as follows:

Algorithm: For deciding greedy degree

```python
# Python3 program to solve fractional
# Knapsack Problem
class ItemValue
"""Item Value DataClass"""
def __init__(self, wt, val, ind):
        self.wt = wt
        self.val = val
        self.ind = ind
        self.cost = val / wt
def __lt__(self, other):
        return self.cost < other.cost
# Greedy ApproachclassFractionalKnapSack:
"""Time Complexity O(n log n)"""
    @staticmethod
defgetMaxValue(wt, val, capacity):
"""function to get maximum value """
        iVal =[]
for i inrange(len(wt)):
            iVal.append(ItemValue(wt[i], val[i], i))
for i inrange(len(wt)):
print("index=",iVal[i].ind,"weight=",iVal[i].wt,"value=",iVal[i].val)
        # sorting items by value
print("After Sorting... ")
        iVal.sort(reverse=True)
        for i in range(len(wt)):
            print("cost =",iVal[i].cost)
Goptp = 0 # optimal value
    Q = 0 # no of items included in solution
        includedWeight= []
        includedVal = []
for i in iVal:
            curWt =int(i.wt)
            curVal =int(i.val)
if capacity - curWt >= 0:
                capacity -= curWt
                Goptp += curVal
                Q += 1
    includedWeight.append(curWt)
        includedVal.append(curVal)
print("selected item ",curVal,curWt)
print("No of Items= m = ",Q
#for i in range(len(includedWeight))
#print("index=",includedTerms[i].ind,"weight=",includedTerms[i].wt,
"value=",includedTerms[i].val)
print("GW= ",includedWeight)
print("Included Values=",includedVal)
        n = len(wt)
while(Q< n//2):
            n = n//2
```

```
    j = inequality2(includedWeight,Goptp,includedVal,Q,0.7,0.5,len(iVal))
return Goptp,Q,j,iVal
definequality2(weights,Goptp,p,Q,lamb,zeta,n):
print('inequality 2 ')
    gw = 0
    for i in weights:
        gw += i
    print(gw)
  for i in range(Q):
        sumOfW = 0
        sumOfp = 0
        for j in range(Q):
            #print(j)
            sumOfW += weights[j]
            sumOfp += p[j]
            if(i==j):
                x1 = sumOfW / gw
                y1 = sumOfp / Goptp
print(x1,y1)
                if(min(x1,y1)<= lamb):
print("first condition become true which is <= lambda")
                    sumOfW += weights[j+1]
                    sumOfp += p[j+1]
                    x2 = sumOfW / gw
                    y2 = sumOfp / gw
if(min(x2,y2)> lamb):
print("second condition become true which is > lambda")
                        sumOfW1,sumOfW2 = 0,0
                        sumOfp1,sumOfp2 = 0,0
for i inrange((len(weights)-1)//2):
                            sumOfW1 += weights[i]
print("step 1",i)
print('j = ',j)
whileTrue:
#code check for large values
for i in`  (weights)-1)//2 )+ 1,j):
                                sumOfW2 += weights[i]
                print("step 2",i)
                                sumOfW2 += weights[j]
print(sumOfW2)
break
  x3 ((sumOfW1 + sumOfW2) / gw
for i inrange((len(weights)-1)//2):
                sumOfp1 += p[i]
whileTrue
for i inrange(((len(weights)-1)//2) + 1,j):
                                sumOfp2 += p[i]
                                sumOfp2 += p[j]
break
                        y3 =(sumOfp1 + sumOfp2) / Goptp
print(x3,y3)
if(min(x3,y3) < zeta):
print("third condition become true ")
print(end='\n')
return j
```

10.3 DYNAMIC EXPECTATION EFFICIENCY (DEE)

DEE model works to choose the item from the unselected (n_m) items known as the candidate region. We have chosen the classical expectation efficiency for the theory proposed by Kahneman (1979) [13], which was further developed by A. Tversky (1992) [14]. This chapter will treat it as a heuristic method to solve the Knapsack 0–1. The aim is to expect the inclusion of the next item by selecting the expected efficiency of the current items. Let E_j denote the expected efficiency of item j and E_j represent the potential of the items j. These can be included in the Knapsack if the item expectance is more efficient than the other expectation efficiency of the items in the remaining items. The scheme of expectation efficiency is shown as follows:

$$E_j = \frac{P_j\left(c - \sum_l^{j-1} w_l\right) w_j}{P_{j-1}\left(c - \sum_l^{j-1} w_l\right) w_{j-1}} = \frac{e_j}{e_{j-1}} \tag{10.6}$$

To illustrate the application of equation (6) and compute the expected efficiency of the items to be included, the Knapsack is specified in Table 10.1. This model seems to fit the Knapsack 0–1 optimal solution factor, which contains the variables n, w, p, c, and j mentioned in equation (1) and is also related to efficiency e, which appears in equation (6). The model has factors associated with updating equation (6), which is mentioned as follows:

$$E(j, w, p, e, n, c) \propto j, w, p, e, n\,c \tag{10.7}$$

The following exercise is established and is a strong model for equation (7) due to abstraction. The remaining capacity of the Knapsack fills progressively by including the items one by one in the Knapsack until it goes to zero in the remaining set. Suppose the item $(j-1)$ is in the Knapsack and unselected capacity of the Knapsack is more significant than zero. In that case, the unselected item's expected profit is to maintain Knapsack's capacity, refer to $e(j-1)$ can be obtained.

Table 10.1 KP-1 Illustration of Computing Greedy Degree

C	n	wi	pi	ei	$z^G W^G$
50	5	{5,15,20,25,30}	{12,30,44,46,50}	{2.4,2.0,1.76,1.70,1.66}	86 45

The optimal profit, in abbreviated form, constitutes as follows:

$$opt\, p = e_{j-1}\left(c -\sum_{l=1}^{m} w_l - \sum_{l=m+1}^{j-1} w_l\, x_l \right) \tag{10.8}$$

Let us assume

$$A = \left(c -\sum_{l=1}^{m} w_l - \sum_{l=m+1}^{j-1} w_l\, x_l \right) \tag{10.9}$$

It is surely p_j is smaller than $opt\, p$ to maintain the differences of p_j and w_j take an average of $opt\, p$ and subtract the p_j the outcome is Δp as follows:

$$\Delta p = e_{j-1}\left(c -\sum_{l=1}^{m} w_l - \sum_{k=m+1}^{j-1} w_l\, x_l \right) - p_j\,(n-j+1) \tag{10.10}$$

$$Ef\left(j, w, p, e, n, c \right) = \frac{e_j\left(c -\sum_{\kappa=1}^{m} w_k - \sum_{\circ=m+1}^{j-1} w_k\, x_k \right) - \left(n-j+1 \right) p_j}{e_{j-1}\left(c -\sum_{\circ=1}^{m} w_k - \sum_{\circ=m+1}^{j-1} w_k\, x_k \right) - \left(n-j+1 \right) w_j} \tag{10.11}$$

$$m+1 \le j \le n$$

Algorithm: For deciding dynamic expectation efficiency (DEE)

```python
def dynamicExpectationEfficiency(wt,p,n,m,cap,ratio):
print("dynamicExpectationEfficiency")
print("Parameters : wt = ",wt,"p = ",p,"n = ",n,"m = ",m,"cap= ",cap)
    listDf = []
    listA = []
for k in range(m,n):
print("Iteration start with value k = ",k)
        wk = 0
#Summation k = 1 to m of Wk
for i in range(m):
            wk += wt[i]
print(wk)
        q = m
        Ai  = 0
 i  = k
        wx = 0
for a in range(q,i+1):
print("Index a = ",a)
        wx += wt[a]
print(end='\n')
print('wx   = ',wx)
        Ai = cap - wk - wx
        listA.append(Ai)
        B = (n - k ) * p[k]
        C = (n- k  ) * wt[k]
        fi = ratio[k].cost / ratio[k-1].cost
#Find DF from Equation
print("Ai = ",Ai,"B = ",B," C = ",C ,  " fi = ",fi);
        Dfi = fi*((Ai * ratio[k-1].cost - B) / (Ai - C) )
```

```
print("Dfi ", Dfi )
        listDf.append(Dfi)
print("Iteration End with value k = ",k)
print(listDf)
print(listA)
return (listDf,listA,m,wt,p)
def removalProcess(listDf,listA,m,wt,p)
print(listDf)
print(listA)
    lengthA = len(listA)
    index = 0
print("length = ",lengthA," index = ",index)
if(lengthA > 0):
while(True):
print("A at index ",index," is ",listA[index])
if(listA[index] < 0):
                dl = min(listDf)
                dlIndex =  listDf.index(min(listDf))
                newIndex = m + dlIndex
print("Removal Process ",dl,dlIndex,newIndex)
del wt[newIndex]
del p[newIndex]
print("Remaining Weights ",wt)
            index += 1
if(index == lengthA):
break
```

10.4 STATIC EXPECTATION EFFICIENCY (SEE)

Static expectation assumption is to expect the value of an economic variable before computing the next period to be included in item values and equal to the current value of the variable obtained nearly optimal solution z^G in the algorithm (1). It is proposed to update the time for the best profit in the SEE model. The range of the best profit is defined in the application of the following theorem:

Theorem: Let $z^U = z^G P_{Q+1}$ and the best profit belonging to $\{z^G, z^U\}$.
Proof: We adopted the reduction of absurdity for this proof.

(i) z^G the objective value of the greedy algorithm is clearly shown $z^U \leq best\ profit$.

(ii) Let $z^U \leq best\ profit$. Then the item $(q + 1)$ should be included in the knapsack after the run algorithm (1) and $z^D = z^G$ gives the contradiction of $z^D = z^G + P_{Q+1}$.

Table 10.2 KP-1 Instance DEE Model

C	n	m	wi	pi	ei	$z^D W^D$
50	5	2	{5,15,20,25,30}	{12,30,44,46,50}	{2.4,2.0,1.76,1.70,1.66}	92 50

(iii) Let $z^U \geq best\ profit$. Then the item $(Q + 1)$ should be included in the knapsack after the run algorithm (1) gives the contradiction of $z^D = z^G + P_{Q+1}$.

A combination of the statements mentioned in (i)–(iii) is proof of Theorem 2.4.1.

In the DEE model, Ae_{j-1} continuously modified in equation (8), when computing the efficiency value is computed by z^D. It is the optimal value of the candidate region and is uncertain and equal to the best profit. So it is required to be updated. Given this motivation to keeping preserve the inflexibility Ae_{j-1} in equation (8), it suggests a static expectation efficiency (SEE) model, which is described as follows:

$$Sf(j,w,p,e,n,c,t) =$$

$$\frac{e_j \left(optp - \sum_{k=1}^{m} w_k - \sum_{k=m+1}^{j-1} w_k x_k \right) - (n-j+1) p_j}{e_{j-1} \left(c - \sum_{k=1}^{m} w_k - \sum_{k=m+1}^{j-1} w_k x_k \right) - (n-j+1) w_j}; \tag{10.12}$$

$$m+1 \leq j \leq n \tag{10.13}$$

$$optp(t) = z^D + t \tag{10.14}$$

Here t is a positive number.

The best profit is in the range of $\{z^D, z^U\}$. According to Theorem 2.4.1. For good efficiency, it should be updated, and the objective function value obtained from the SEE model is not less than the DEE model objective function. On this basis, we imposed the constraint as follows:

$$\begin{cases} z^D \leq optp(t) < z^U, z^D < z^G \\ z^G \leq optp(t) < z^U, z^D \geq z^G \end{cases} \tag{10.15}$$

Combine equations (12) and (13) as follows:

$$\begin{cases} 0 \leq t < z^U - z^D, z^D < z^G \\ z^G - z^D \leq t < z^U - z^D, z^D \geq z^G \quad t \in z^+ \end{cases} \tag{10.16}$$

If t keeps unchanged, there is a certainty in the item selections, and the optimality $otp\ p(t)$ is the obtained objective function value. The new objective function value z^{NS} obtained by the SEE model by not considering the $(n-m)$ items. The SEE model objective function value can be obtained from the specifications $sf(m+1), sf(m+2)\ldots\ldots\ldots sf(n)$ and z^{NS} is obtained in the algorithm mentioned in Section 2.4.2. Let us consider

$$T = optp - \sum_{\circ=1}^{m} w_k - \sum_{k=m+1}^{j-1} w_k x_k) \tag{10.17}$$

The objective function value obtained by using the algorithm mentioned in 2.4.2 is to replace the objective function value of the DEE model.

Algorithm: Static expectation efficiency (SEE)

```
def staticExpectationEfficiency(wt,p,n,m,cap, ratio, optp):
print("staticExpectationEfficiency")
print("Parameters : wt = ",wt,"p = ",p,"n = ",n,"m = ",m,"cap= ",cap)
    listSf = []
    listT = []
for k inrange(m,n):
print("Iteration start with value k = ",k)
        wk = 0
#Summation k = 1 to m of Wk
        for i in range(m):
            wk += wt[i]
print(wk)
        q = m
        Ai = 0
i = k
        wx = 0
for a inrange(q,i+1):
print("Index a = ",a)
            wx += wt[a]
print(end='\n')
print('wx = ',wx)
        Ai = cap - wk - wx
 B =(n - k ) * p[k]
        C =(n- k   ) * wt[k]
        fi = ratio[k].cost / ratio[k-1].cost
        T = optp - wk - wx
        listT.append(T)
        #Find DF from Equation
print("Ai   ",Ai,"B = ",B," C = ",C , " fi = ",fi , "Ti = ",T);
        Sfi = fi*( (T - B) / (Ai - C))
print("Sfi ", Sfi )
        listSf.append(Sfi)
print("Iteration End with value k = ",k)
print(listSf)
print(listT)
    return (listSf,listT,m,wt,p)
```

10.4.1 Parallel computing

The new objective function value depends on t. As t changes, the objective functions also change. Let N be the number of objective function values.

Generally, serial processors consume time to compute the new objective function values. We follow the parallel computing method to reduce the computing time while using a computer program. In Algorithm 3, the objective function is updated to $Z^{NS}(t)$ by the different parallel processors. The updated objective value in serial takes N-steps, and parallel takes one step. The process of updating objective function value by serial and parallel methods is shown in Figure 10.1.

If $z^D < z^G$, then N new objective values are $z^{NS}(0), z^{NS}(1), \ldots \ldots \ldots z^{NS}(z^U - z^D - 1)$. The best profit denoted by z^B is obtained by equation (17).

Let z^B be the best profit, which can be obtained as follows:
If $z^D < z^G$ then

$$z^B = \max\{z^D; u / u \in \{z^{NS}, 0 \le t < z^U - z^D\} \tag{10.18}$$

Serial method

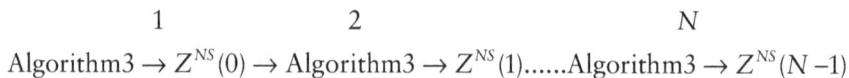

$$\underset{1}{\text{Algorithm3}} \to Z^{NS}(0) \to \underset{2}{\text{Algorithm3}} \to Z^{NS}(1)......\underset{N}{\text{Algorithm3}} \to Z^{NS}(N-1)$$

Parallel method

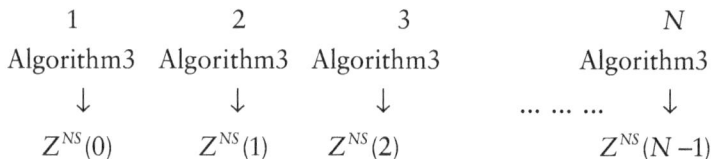

1	2	3	N
Algorithm3	Algorithm3	Algorithm3	Algorithm3
↓	↓	↓	... ... ↓
$Z^{NS}(0)$	$Z^{NS}(1)$	$Z^{NS}(2)$	$Z^{NS}(N-1)$

Figure 10.1 The serial method with one output and parallel methods with n output processing and updating the objective function values in case of static expectation efficiency running the algorithm.

If $z^D \geq z^G$ then

$$z^B = \max\{z^D; u \,/\, u \in \left\{z^{NS}, 0 \leq t < z^U - z^G\right\} \tag{10.19}$$

Suppose lu item can be included by updating the candidate objective value in Knapsack. Then the region for the content item $(lu + m)$ is concluded.

In Figure 10.2, particular first n items are rearranged with efficiency and sorted them in descending order. Second, the greedy degree model proposed the first m items included in the Knapsack and never removed from the Knapsack. Third, the DEE model used to include an item from the unselected item $(n - m)$ into Knapsack, and the last unselected item $(n - m - u)$ is regarded as unknown item regions. Fourthly, the DEE model with parallel computing method is used to update candidate objective value, and lu items are included in the Knapsack. At last, the first m items are determined and lu items constitute finally specific region.

Algorithm: GDSEE algorithm

```
Input: wi,pi,n and c
Output: z^B
        Run greedy algorithm 1
        Obtain:m
        Run dynamic expectation efficiency algorithm 2
        Obtain:z^D
Run Static expectation efficiency algorithm 3:By parallel method
If z^D < z^G thenz^B = max {z^D; u/u ∈ {z^NS, 0 ≤ t < z^U − z^D}
        else
                    z^B = max {z^D; u/u ∈ {z^NS, 0 ≤ t < z^U − z^G}
        return
```

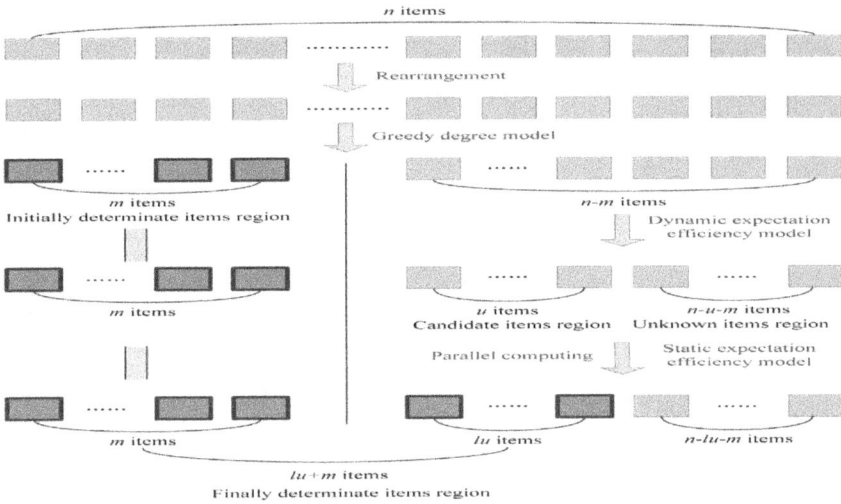

Figure 10.2 Process of GDSEE for solving KP-1 [10].

10.5 TIME COMPLEXITY DESCRIPTION

Theorem: Greedy algorithm runs in $O(n)$.

Proof. According to the greedy algorithm, the running check has three essential steps. At first, it checks whether the constraint inequality $Q < n/2$ is met or not; in the second part, it prevents the inequality (2) of the constraints met or not, and in the third part, checking, it checks both and runtime complexity is $O(n)$. In the second part algorithm, search items $n/2$ are in the beginning, and thereafter, the searching items are $n/4$ and so on. At the end of the search, there are $\dfrac{n}{2^k}$ items. The *kth* time search for the lower limit of the inequality is shown as

$$1 \le \frac{n}{2^k} \le 2 \tag{10.20}$$

$$logn - 1 < k \le log2^n \tag{10.21}$$

Since the third part of the algorithm runs in order $O(logn)$, that is, $O(n)$.

Theorem: DEE algorithm run time complexity is $0(n)$.

Proof. The runtime during the execution of equation (10) in the DEE model is $(n - m)$, and thus, this run in $0(n)$ and the remaining part run in the same fashion in $0(n)$. DEE model is also run in $0(n)$.

Theorem: SEE algorithm runtime order is $0(n)$.

Proof. DEE algorithm proof is the same as the SEE algorithm. The principle of both algorithms is the same. Hence, the SEE algorithm also runs in $O(n)$

Theorem: GDSEE algorithm runtime order is $O(n)$.

Proof. GDSEE algorithm is a combination of many parts; there are four prime constituent parts. The first part is the greedy degree, the second part is the DEE algorithm, the third part is SEE, and the fourth part is the computation of the best profit as z^B, which is computed by equations (17) and (18). Runtime complexity is $O(n)$. As the fourth part, search for the best profit in the search $(z^U - z^D)$ or $(z^U - z^G)$ element when Z^B is selected as the best profit is run in $O(n)$. Hence GDSEE is run over in $O(n)$

10.6 SPACE COMPLEXITY

Theorem: Space complexity of GDSEE is in $O(n^2)$.

Proof. In the DEE algorithm, there are $(n-m)$ items that obtain objective function value z^D. For static expectation, there are $(n-m)$ items; it is required to obtain N new objective function value z^{NS} in parallel methods. The total $N*(n-m)$ static expectation efficiency values come out in the process, which needs to store in space with $N*(n-m)$ units. Hence, the space complexity of GDSEE is getting of $O(n^2)$.

10.7 EXPERIMENTAL RESULTS AND DATA ANALYSIS

The experimental studies consist of the running time of the programming code with the best profit, worst profit, and the storage space's size as the performance valuation measure. The correctness, feasibility, effectiveness, and stability are essential for evaluating the performance of the greedy degree expectance efficiency GDSEE. Four groups of the simulation experiment according to 50 Knapsack 0–1 instances are presented in the experimental results. First, a numerical instance is given to illustrate the computational process and correctness of greedy degree evaluation efficiency. Second, ten cases are tested to demonstrate the feasibility of greedy degree evaluation efficiency. Third, the greedy degree evaluation efficiency is compared with the data set for the chemical reaction algorithm with the greedy strategy mentioned in the work of K. T. Truong [15], which is denoted as CROG. In the fourth experiment, the greedy degree evaluation efficiency is computed in different test case libraries with 44 instances to reveal the stability of the greedy degree evaluation effectiveness. The simulation test is done using the software Python version-3.

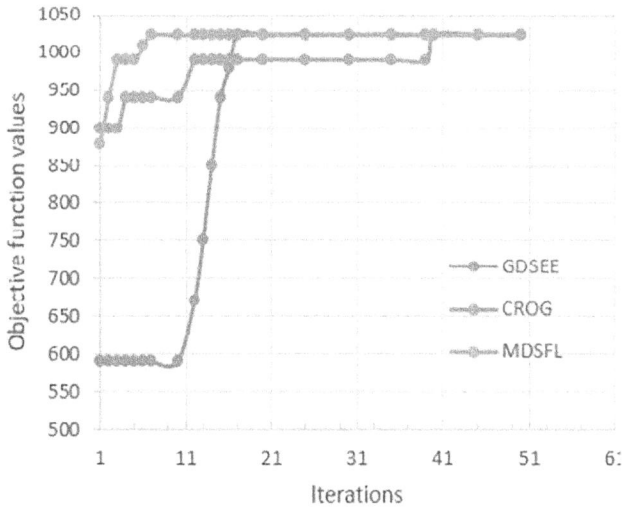

Figure 10.3 Detailed changing process of objective function value for f2.

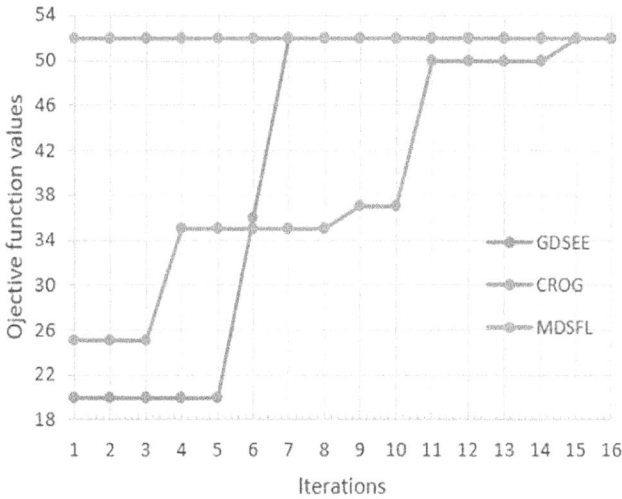

Figure 10.4 Detailed changing process of objective for f4.

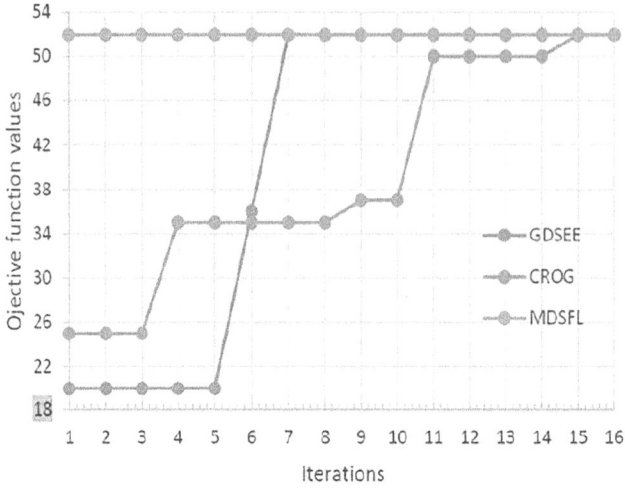

Figure 10.5 Detailed changing process of objective function value for f6.

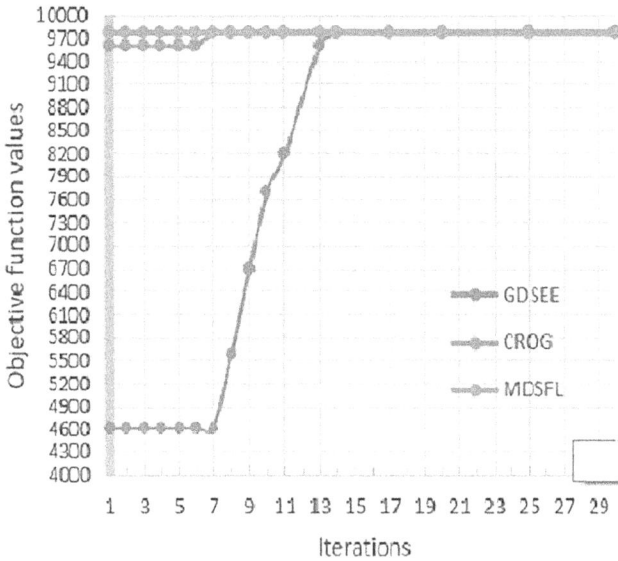

Figure 10.6 Detailed changing process of objective for f8.

10.7.1 Simulation test

Numerical instance for correctness test

Give one instance KP2: $c = 620$, $n = 20$, $W = (22, 36, 32, 18, 35, 26, 44,$
$50, 45, 44, 48, 50, 12, 52, 24, 52, 60, 28, 55, 38)$, and $P = (56, 60, 48,$

25, 72, 40, 55, 55, 50, 37, 30, 48, 18, 78, 30, 60, 45, 35, 70, 48). The process of obtaining the best profit is presented as follows.

Step 1: Rearrange 20 items. W = (22, 35, 36, 26, 12, 32, 52, 18, 55, 38, 28, 24, 44, 52, 45, 50, 50, 44, 60, 48), and P = (56, 72, 60, 40, 18, 48, 78, 25, 70, 48, 35, 30, 55, 60, 50, 55, 48, 37, 45, 30).

Step 2: Run greedy Algorithm 1. Z^G 848, $W^G = 619$, $Q = 17$, and $m = 10$.

Step 3: Run DEE Algorithm 2. $Z^D = 848$.

Step 4: Compute the range of the best profit according to Theorem 1. $Z^U = 885$ and $Z^D = Z^G = 848$, and then optp(t) $\in$ [848, 885).

Step 5: Obtain some new objective function values by parallel computing method. Let $opt\,p(t)$ be substituted into equation (10), and 37 new objective function values are 848, 849, . . . , 884, respectively.

Step 6: Obtain the best profit by equation (17) since $Z^D = Z^G = 848$ in Step 4. $Z^B = \max\{848, Z^{NS}(0), Z^{NS}(1), Z^{NS}(36)\} = 848$. For KP2, the best profit is 848, and the detailed changing process of objective function value is depicted in Figure 10.7.

10.7.2 Feasibility test

Ten instances for standard test case libraries of Knapsack 0–1 are considered to determine the feasibility of greedy degree evaluation effectiveness. The size of the items are 5, 8, 10, 20, 23, 50, 80, 100, 100, and 200. The experimental results are composed of five indexes, that is, parallel update,

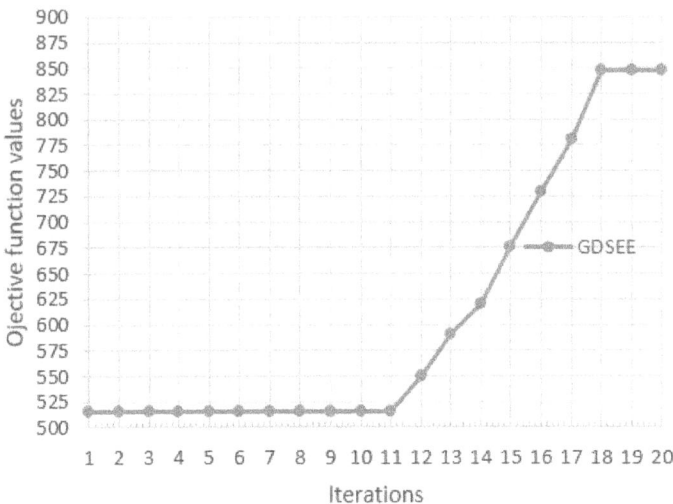

Figure 10.7 Detailed changing process of objective function value.

Table 10.3 Algorithm Test

Instance	Item	Size	Capacity (c)	Best Profit z^B	Worst Profit z^B	NPU Running Times
1.	5	50	92	86	46	25.366
2.	8	200	390	390	33	25.362
3.	10	269	295	294	30	27.527
4.	20	879	1024	1018	245	29.472
5.	20	620	848	848	37	28.468
6.	23	10000	9767	9767	972	29.213
7.	50	300	1063	1060	11	30.121
8.	80	800	2085	2085	112	30.748
9.	100	1000	2617	2610	43	32.412
10.	100	1000	2614	2613	24	31.962
11.	200	2000	5185	5185	26	38.457

best profit, worst profit, size of storage space, and running time. They are shown in Table 10.3.

Several useful conclusions are presented in Table 10.3, namely, (a) GDSEE can solve Knapsack 0–1 and obtain the best profit; (b) it requires many storage-space greedy degree expectation efficiency values; (c) the running times increase with the increasing of numbers of items; (d) a number of items are not proportional to the size of storage space, since it is determined by the upper bound of the total profit concerned to any numbers of the items; (e) for some knapsack 0–1, there exist several items that are the same but with different best profit since the best profit is determined by six factors as mention in equation (7).

10.7.3 Comparison among the algorithms

We have compared the three algorithms namely greedy degree evaluation efficiency, chemical reaction optimization with greedy strategy [15], and modified discrete shuffled frog leaping algorithm [16] according to the effectiveness and stability factors.

10.7.4 Effectiveness test

Table 10.4 compares the results of this algorithm's running times based on the best profit, worst profit, difference between the best and worst profit, population size, and numerous simultaneous updates. The detail of changing process of the values for f2, f4, f6, f8, and f10 is shown in Figures 10.3–10.6. In Table 10.4, the chemical reaction optimization greedy algorithm obtains the best profit by setting the population size to 20. In contrast, the modified MDSF algorithm obtains the best profit by setting the population size for

Table 10.4 Algorithm Test

Instances	Algorithms	Best Profit z^B	Worst Profit z^B	Population Size	Running Times
f2	CROG	1024	1018	20	36.100
	MDSFL	1024	1018	200	38.653
	GDSEE	1024	1018	11	29.475
f4	CROG	23	16	20	29.714
	MDSFL	23	23	200	19.360
	GDSEE	23	22	8	26.201
f6	CROG	52	50	20	30.917
	MDSFL	52	52	200	20.481
	GDSEE	52	52	15	29.472
f8	CROG	9767	9765	20	34.605
	MDSFL	9767	9767	200	31.555
	GDSEE	9767	9767	486	29.213

200, and the greedy degree effectiveness efficiency obtains the best profit for numbers of parallel updates for 11, 8, 10, 486, and 11, respectively.

10.7.5 Stability test

In the context of the stability of greedy-degree effective efficiency, three standard case libraries, namely chemical reaction optimization greedy algorithm and modified discrete shuffled frog leaping greedy degree effectiveness algorithm, are compared with 10, 20, and 14 instances. The average running time is corrected with the Knapsack 0–1, and solvability determination is considered the stability test.

10.8 CONCLUSIONS AND FUTURE WORK

Knapsack 0–1 has been widely applied in real-world modeled optimization problem solving such as capital budgeting resources allocation, Portfolio selection, project investment, and decision-making. This chapter describes a hybrid heuristic algorithm based on greedy degree expectation efficiency for solving the Knapsack 0–1 problem. In the projected greedy degree expectation efficiency, the greedy degree model is presented in the Python-coded form, which puts some items into Knapsack in the beginning. Furthermore, two heuristic expectation efficiency models are designed to generate the objective function values from the remaining items. The parallel computing methods are introduced to accelerate the iterated values of the objective function. The time complexity of the greedy degree expectation efficiency is analyzed, and its run time is $O(n)$, which is appreciable. The space complexity of greedy degree efficiency is also analyzed, and it's run in $O(n^2)$, which is acceptable due to the cheap memory cell availability. The performance of greedy degree effectiveness efficiency is investigated through four groups of experiments. The expectation efficiency models are encrypted in the Python codes.

There is a close relationship with the Knapsack 0–1 in the CPU and data bus packing bin during the identity and excess management process, which is the future work and the application of Knapsack 0–1 in computer process operating system modeling.

REFERENCES

[1] S. Martello, D. Pisinger, P. Toth, New trends in exact algorithms for the 0–1 knapsack problem, *Eur. J. Oper. Res.* 123 (2000) 325–332.

[2] P.J. Kolesar, A branch and bound algorithm for the knapsack problem, *Manage. Sci.* 13 (1967) 723–735.

[3] E. Horowitz, S. Sahni, Computing partitions with applications to the knapsack problem, *J. Assoc. Comput. Mach.* 21 (1974) 277–292.

[4] D. Pisinger, An expanding-core algorithm for the exact 0–1 knapsack problem, *Eur. J. Oper. Res.* 87 (1995) 175–187.

[5] S. Martello, P. Toth, A bound and bound algorithm for the zero-one multiple knapsack problem, *Discret. Appl. Math.* 3 (1981) 275–288.

[6] P. Toth, Dynamic programming algorithm for zero-one knapsack problem, *Comput.* 25 (1980) 29–45.

[7] A. Rong, J.R. Figueira, Computational performance of basic state reduction based on dynamic programming algorithms for bi-objective 0–1 knapsack problems, *Comput. Math. Appl.* 63 (2012) 1462–1480.

[8] J.R. Figueira, L. Paquete, M. Simoes, Algorithm improvements on dynamic programming for the bi-objective 0–1 knapsack problem, *Comput. Optim. Appl.* 56(2013) 97–111.

[9] P. Vasant, N. Barsoum, J. Webb, *Innovation in Power, Control, and Optimization: Emerging Energy Technologies*, IGI Global, 2011.

[10] Jianhui Lv, Xingwei Wang, Min Huang, Hui Cheng, Fuliang Li, Solving 0–1 knapsack problem by greedy degree and expectation efficiency, *Appl. Soft Comp.* 41 (2016) 94–103.

[11] R. Merkle, M. Hellman, Hiding information and signatures in trapdoor knapsacks, *IEEE Trans. Inf. Theory* 24 (1978) 525–530.

[12] G. Mavrotas, D. Diakoulaki, A. Kourentzis, Selection among ranked projects under segmentation, policy and logical constraints, *Eur. J. Oper. Res.* 187 (2008) 177–192.

[13] D. Kahneman, A. Tversky, Prospect theory: An analysis of decision under risk, *Econometrica.* 47 (1979) 263–291.

[14] A. Tversky, D. Kahneman, Advances in prospect theory: Cumulative representation of uncertainty, *J. Risk Uncertain.* 5 (1992) 297–323.

[15] K.T. Truong, K. Li, Y. Xu, Chemical reaction optimization with greedy strategy for the 0–1 knapsack problem, *Appl. Soft Comput.* 13 (2013) 1774–1780.

[16] K.K. Bhattacharjee, S.P. Sarmah, Shuffled frog leaping algorithm and its application to 0/1 knapsack problem, *Appl. Soft Comput.* 19 (2014) 252–263.

Chapter 11

A transportation model with rough cost, demand, and supply

Subhakanta Dash and S. P. Mohanty

11.1 INTRODUCTION

The transportation problem involves the distribution of goods or services from a set of sources to a set of destinations through a network. There are different routes and different transportation costs for the routes. The transportation problem aims to determine the number of units of goods or services to be transported so that all demands and supplies are satisfied with the minimum transportation cost. The traditional transportation problem consists of one objective function and two constraints, namely, source constraints with the supply and destination constraints with the demand, which was first initiated by Hitchcock (1941) and later developed by Koopmans (1947). Dantzig (1951) proposed the simplex method and applied it to solve transportation problems as a linear programming problem. Since then, many researchers have developed many algorithms to solve transportation problems. In traditional transportation problems, the parameters like unit transportation cost, demand at destinations, and supply at sources are taken as deterministic values. Due to various complexities in the real world, such as unpredictable weather conditions, road conditions, traffic conditions in the road, change in the sale, and change in the attitude of the customers, it is not appropriate to regard the unit transportation cost, the supplies, and the demands as deterministic. They should be considered as variables. Williams (1963) has developed a stochastic model of transportation problems considering the parameters as random variables. Since then, many researchers have studied stochastic models of transportation problems. Later it was observed that the input data are often imprecise owing to incomplete or unobtainable information. So, researchers tried to consider the parameters as fuzzy variables. Chanas et al. (1984) presented a fuzzy linear programming model to solve transportation problems with crisp cost, fuzzy supply, and fuzzy demand.

Furthermore, it is observed that, in many situations, no investigated data are available to estimate the appropriate probability distribution of the assumed random variables. In this situation, some domain experts are invited to give their subjective estimates of the above parameters. To deal

DOI: 10.1201/9781003462422-11

with human uncertainty, the rough set theory developed by Pawlak (1982) and Liu (2004, 2007) developed the uncertainty theory, which has become a powerful tool for dealing with human belief degree. Using uncertain variables, some authors have developed models for transportation problems. Guo et al. (2015) have developed a transportation model considering the supply as a random variable and the cost and the demand as uncertain variables. Yuhong Shang and Kai Yao (2012) have developed the transportation model considering the cost, supply, and demand as uncertain variables. Kundu et al. (2015) have developed a solid transportation model with product blending and parameters as the rough variable.

Practically, it is very likely that the experts' subjective estimates of the parameters are given in a certain range of values that rough variables can characterize. In this chapter, a transportation model is developed considering the unit cost of transportation, the supply, and the demand as rough variables.

The rest of this chapter is organized as follows. Section 11.2 presents some basic concepts of rough variables and their properties. In Section 11.3, a transportation model with rough cost, demand, and supply is developed. One numerical example is given in Section 11.4 to illustrate the model. Finally, the conclusion is given in Section 11.5.

11.2 PRELIMINARIES

This section presents some concepts and notions of the rough variable. The concept of rough variable is introduced by Liu (2004) as the uncertain variable. The following definitions are based on Liu (2004, 2007).

Definition 1: Let Λ be a nonempty set, A be σ-algebra of subsets of Λ, Δ be an element in A, and π be a nonnegative, real-valued, additive set function on A. The quadruple $(\Lambda, \Delta, A, \pi)$ is called a rough space.

Definition 2: A rough variable ξ on the rough space $(\Lambda, \Delta, A, \pi)$ is a measurable function from Λ to the set of real numbers $\Re$ such that for every Borel set B of $\Re$, we have $\{\lambda \in \Lambda \mid \xi(\lambda) \in B\} \in A$.

Then the lower and upper approximations of the rough variable ξ are defined as follows

$$\overline{\zeta} = \{\zeta(\lambda) | \lambda \in \Lambda\} \text{ (Upper approximation)}$$

$$\underline{\zeta} = \{\zeta(\lambda) | \lambda \in \Delta\} \text{ (Lower approximation)}$$

Definition 3: ([a, b], [c, d]) with $c \leq a < b \leq d$ is a rough variable, where $\xi(\lambda) = \lambda$ from the rough space to the set of real numbers and $\Lambda = \{\lambda \mid c \leq \lambda \leq d\}$ and $\Delta = \{\lambda \mid a \leq \lambda \leq b\}$. A is the Borel algebra on Λ; π is the Lebesgue measure.

Definition 4: Let $((\Lambda, \Delta, A, \pi))$ be a rough space. Then the upper and lower trust of event A is defined by $\underline{Tr}(A) = \dfrac{\pi\{A\}}{\pi\{\Lambda\}}$ and $\overline{Tr}(A) = \dfrac{\pi\{A \cap \Delta\}}{\pi\{\Lambda\}}$

The trust of event A is defined as

$$Tr(A) = \frac{1}{2}\left(\underline{Tr}(A) + T\,\overline{r}(A)\right).$$

Definition 4: Let ξ be rough variables defined on the rough space $(\Lambda, \Delta, A, \pi)$ and $\alpha \in (0,1]$ then $\xi_{sup}(\alpha) = \sup\{r \mid Tr\{\xi \geq r\} \geq \alpha\}$ is called α-optimistic value of ξ.

$\xi_{inf}(\alpha) = \inf\{r \mid Tr\{\xi \leq r\} \geq \alpha\}$ is called α-pessimistic value of ξ.

Definition 5: Let ξ be a rough variable. Then the expected value of ξ is defined by

$$E[\xi] = \int_{0}^{+\infty} Tr\{\xi \geq r\}dr - \int_{-\infty}^{0} Tr\{\xi \leq r\}dr$$

provided that at least one of the two integrals is finite.

Definition 6: The trust distribution $\phi: [-\infty, \infty] \rightarrow [0, 1]$ of a rough variable ξ is defined by

$$\Phi(x) = Tr\,\{\lambda \in \Lambda \mid \zeta(\lambda) \leq x\}$$

Definition 7: The trust density function $f : R \rightarrow [0,\infty)$ of a rough variable ξ is a function such that

$$f(x) = \int_{-\infty}^{\infty} \phi(y)dy \text{ holds for all } x \in (-\infty, \infty), \text{ where } \phi \text{ is trust distribution}$$
of ξ.

Definition 8: If $\xi = ([a,b],[c,d])$ is a rough variable such that $c \leq a < b \leq d$, then the trust distribution $\phi(x) = Tr\{\xi \leq x\}$ is

$$\phi(x) = \begin{cases} 0 & \text{if } x \leq c \\[2mm] \dfrac{x-c}{2(d-c)} & \text{if } c \leq x \leq a \\[2mm] \dfrac{[(b-a)+(d-c)]x + 2ac-ad-bc}{2(b-a)(d-c)} & \text{if } a \leq x \leq b \\[2mm] \dfrac{x+d-2c}{2(d-c)} & \text{if } b \leq x \leq d \\[2mm] 1 & \text{if } x \geq d \end{cases}$$

and the trust density function is defined as

$$
f(x) = \begin{cases}
\dfrac{1}{2(d-c)} & \text{if } c \le x \le a \ \text{ or } \ b \le x \le d \\[3mm]
\dfrac{1}{2(b-c)} + \dfrac{1}{2(d-c)} & \text{if } a \le x \le b \\[3mm]
0 & \text{otherwise}
\end{cases}
$$

Definition 9: For a given value r and $\xi = ([a,b],[c,d])$ trust of rough events characterized by $\xi \le r$ and $\xi \ge r$ is, respectively, presented by the following expressions (Liu 2004):

$$
\text{Tr}\{\xi \ge r\} = \begin{cases}
0, & \text{if } r \le d, \\[2mm]
\dfrac{d-r}{2(d-c)}, & \text{if } b \le r \le d \\[3mm]
\dfrac{1}{2}\left(\dfrac{d-r}{d-c} + \dfrac{b-r}{b-a} \right), & \text{if } a \le r \le b \\[3mm]
\dfrac{1}{2}\left(\dfrac{d-r}{d-c} + 1 \right), & \text{if } c \le r \le a \\[3mm]
1, & \text{if } \ r \le c
\end{cases}
$$

$$
\text{Tr}\{\xi \le r\} = \begin{cases}
0, & \text{if } r \le c, \\[2mm]
\dfrac{r-c}{2(d-c)}, & \text{if } c \le r \le a \\[3mm]
\dfrac{1}{2}\left(\dfrac{r-a}{b-a} + \dfrac{r-c}{d-c} \right), & \text{if } a \le r \le b \\[3mm]
\dfrac{1}{2}\left(\dfrac{r-c}{d-c} + 1 \right), & \text{if } b \le r \le d \\[3mm]
1, & \text{if } \ r \ge d
\end{cases}
$$

and

Definition 10: Let ξ be a rough variable in rough space $(\Lambda, \Delta, A, \pi)$. Then α-optimistic value to $\xi = ([a,b],[c,d])$ is

$$
\xi_{\sup}(\alpha) = \begin{cases}
(1-2\alpha)d + 2\alpha c, & \text{if } \alpha \le \dfrac{d-b}{2(d-c)}; \\[3mm]
2(1-\alpha)d + (2\alpha-1)c, & \text{if } \alpha \ge \dfrac{2d-a-c}{2(d-c)}; \\[3mm]
\dfrac{d(b-a) + b(d-c) - 2\alpha(b-a)(d-c)}{(b-a)+(d-c)}, & \text{otherwise.}
\end{cases}
$$

α-pessimistic value to $\xi = ([a,b],[c,d])$ is

$$
\xi_{\inf}(\alpha) = \begin{cases}
(1-2\alpha)c + 2\alpha d, & \text{if } \alpha \le \dfrac{a-c}{2(d-c)}; \\[3mm]
2(1-\alpha)c + (2\alpha-1)d, & \text{if } \alpha \ge \dfrac{b+d-2c}{2(d-c)}; \\[3mm]
\dfrac{c(b-a) + a(d-c) + 2\alpha(b-a)(d-c)}{(b-a)+(d-c)}, & \text{otherwise.}
\end{cases}
$$

Definition 11: Let ξ be a rough variable whose trust density function f exists. If the Lebesgue integral $\int_{-\infty}^{\infty} xf(x)dx$ is finite, then the expected value of ξ is defined as

$$
E(x) = \int_{-\infty}^{\infty} xf(x)dx
$$

11.3 PROBLEM DESCRIPTION

In reality, transportation planning is made in advance sometimes. Due to many uncertain factors like weather conditions, road conditions, and changes in sales due to customers' attitudes, the supply, demand and cost of transportation may not be fixed but are somewhat uncertain. Previously, the parameters were taken as random or fuzzy variables to deal with uncertainty. Sometimes considering the parameters as random variables is not ideal due to the absence of sample observations, which is practically happening in many situations based on the degree of belief of subject experts. To deal with these uncertainties, in this chapter, a new approach has been presented by considering the parameters such as the unit cost, supply, and demands, which are all assumed to be rough variables characterized by the subjective judgment of domain experts. Furthermore, like fuzzy variables, the rough variables do not possess any membership function, so it gives an advantage by considering the parameter as a rough variable. Based on this, an uncertain model has been developed.

11.3.1 Deterministic transportation model

Suppose that there are m sources and n destinations in a transportation problem. Let c_{ij} denotes the cost of transporting one unit from source i to destination j and x_{ij} denotes the amount transported from source x_{ij} to destination $j, i = 1,2,\cdots m, \& j = 1,2,\cdots,n$. Then the objective of the problem is to make a transportation plan so that the total transportation cost is minimized.

Let a_i denotes the availability of source i, and b_j denotes the requirement at destination j. Then the transportation problem can be described as

$$
\left\{
\begin{aligned}
& \text{Minimize} \quad \sum_{i=1}^{m}\sum_{j=1}^{n} c_{ij} x_{ij} \\
& \text{Subject to} \\
& \quad \sum_{j=1}^{n} x_{ij} \le a_i, \quad i = 1, 2, ..., m \\
& \quad \sum_{i=1}^{m} x_{ij} \ge b_j, \quad j = 1, 2, ..., n \\
& \quad x_{ij} \ge 0, \quad i = 1, 2, ..., m \text{ and } j = 1, 2, ..., n
\end{aligned}
\right.
\tag{11.1}
$$

The above model is deterministic if the quantities c_{ij}, a_i, b_j are all assumed to be crisp numbers. There are many standard techniques to solve the model.

11.3.2 Uncertain transportation model

In reality, transportation planning is made in advance sometimes. Due to many uncertain factors like weather conditions, road conditions, and changes in sales due to customers' attitudes, the supply, demand and cost of transportation may not be fixed but rather uncertain. To deal with these uncertainties, the quantities c_{ij}, a_i, and b_j are all assumed to be rough variables characterized by domain experts' subjective judgment. Then, model (1) becomes conceptual as in the absence of the natural order of rough variables, the objective function defined in (1) becomes invalid. Thus, the expected value criterion for the objective function and confidence level on the constraint functions are taken. Then we obtain the equivalent form of the model (1) as

$$
\left\{
\begin{aligned}
& \text{Minimize} \quad E\left[\sum_{i=1}^{m}\sum_{j=1}^{n} \tilde{c}_{ij} x_{ij}\right] \\
& \text{Subject to} \\
& \quad Tr\left[\sum_{j=1}^{n} x_{ij} \le \xi_i\right] \ge \alpha_i, \quad i = 1, 2, ..., m \\
& \quad Tr\left[\sum_{i=1}^{m} x_{ij} \ge \eta_j\right] \ge \text{and } \beta_j, \quad j = 1, 2, ..., n \\
& \quad x_{ij} \ge 0, \quad i = 1, 2, ..., m; j = 1, 2, ..., n
\end{aligned}
\right.
\tag{11.2}
$$

where α_i and β_j are some predetermined confidence levels for $i = 1, 2, \dots m, j = 1, 2, \dots, n$.

$$\tilde{c}_{ij} = \left(\left[a_{ij}, b_{ij} \right] \left[c_{ij}, d_{ij} \right] \right), \ c_{ij} \le a_{ij} < b_{ij} \le d_{ij}$$

$$\xi_i = \left(\left[a_i, b_i \right] \left[c_i, d_i \right] \right), \ c_i \le a_i < b_i \le d_i$$

$$\eta_j = \left(\left[a_j^1, b_j^1 \right] \left[c_j^1, d_j^1 \right] \right), \ c_j^1 \le a_j^1 < b_j^1 \le d_j^1$$

Model (2) can be solved, once it is converted into its crisp equivalent form. The following theorems are proposed for the conversion to a crisp identical format.

Theorem 1: Let $\tilde{c}_{ij}$ be rough variables having trust density function $f_{ij}(y)$, then

$$E\left[\sum_{i=1}^{m} \sum_{j=1}^{n} \tilde{c}_{ij} x_{ij} \right] = \frac{1}{4} \sum_{i=1}^{m} \sum_{j=1}^{n} x_{ij} \left(a_{ij} + b_{ij} + c_{ij} + d_{ij} \right)$$

Proof: It follows from the linearity of the expected value operator

$$E\left[\sum_{i=1}^{m} \sum_{j=1}^{n} \tilde{c}_{ij} x_{ij} \right] = \sum_{i=1}^{m} \sum_{j=1}^{n} x_{ij} E\left[\tilde{c}_{ij} \right]$$

For rough independent variables $\tilde{c}_{ij}$, $i = 1, 2, \dots m, j = 1, 2, \dots n$.

$$E\left[\tilde{c}_{ij} \right] = \int_{-\infty}^{\infty} y f_{ij}(y) \, dy$$

$$E\left[\sum_{i=1}^{m} \sum_{j=1}^{n} \tilde{c}_{ij} x_{ij} \right] = \sum_{i=1}^{m} \sum_{j=1}^{n} x_{ij} E\left(\tilde{c}_{ij} \right)$$

$$= \sum_{i=1}^{m} \sum_{j=1}^{n} x_{ij} \left(\int_{-\infty}^{\infty} y f_{ij}(y) \, dy \right)$$

$$= \sum_{i=1}^{m} \sum_{j=1}^{n} x_{ij} \left(\begin{array}{c} \int_{c_{ij}}^{a_{ij}} \dfrac{y}{2\left(d_{ij} - c_{ij} \right)} \, dy \\[2mm] + \int_{a_{ij}}^{b_{ij}} \left[\dfrac{1}{2\left(b_{ij} - a_{ij} \right)} + \dfrac{1}{2\left(d_{ij} - c_{ij} \right)} \right] y \, dy \\[2mm] + \int_{b_{ij}}^{d_{ij}} \dfrac{y}{2\left(d_{ij} - c_{ij} \right)} \, dy \end{array} \right)$$

$$= \frac{1}{4} \sum_{i=1}^{m} \sum_{j=1}^{n} x_{ij} \left(a_{ij} + b_{ij} + c_{ij} + d_{ij} \right)$$

Theorem 2:

If $\xi = ([a,b],[c,d])$ be a rough variable with $c \le a < b \le d$, then for any predetermined confidence level α, $\text{Tr}\{\xi \ge r\} \ge \alpha$, for $0 < \alpha \le 1$ is equivalent to

(i) $(1-2\alpha)d + 2\alpha c \ge r,$ if $\alpha \le \dfrac{d-b}{2(d-c)}$;

(ii) $2(1-\alpha)d + (2\alpha-1)c \ge r,$ if $\alpha \ge \dfrac{2d-a-c}{2(d-c)}$;

(iii) $\dfrac{d(b-a) + b(d-c) - 2\alpha(b-a)(d-c)}{(b-a)+(d-c)} \ge r,$ otherwise.

Proof:

Case 1: For $b \le r \le d$, from definition (9)

$$\text{Tr}\{\xi \ge r\} \ge \alpha \Rightarrow \frac{d-r}{2(d-c)} \ge \alpha$$

$$\Rightarrow d - r \ge 2d\alpha - 2c\alpha$$

$$\Rightarrow d(1-2\alpha) + 2c\alpha \ge r$$

Here the maximum possible value of $\text{Tr}\{\xi \ge r\} \ge \alpha$ will be 0 and minimum value will be $\dfrac{d-b}{2(d-c)}$. So in this case, the value of α must be less than equal to $\dfrac{d-b}{2(d-c)}$.

Case 2: For $c \le r \le a$, from definition (9),

$$\text{Tr}\{\xi \ge r\} \ge \alpha \Rightarrow \frac{1}{2}\left(\frac{d-r}{d-c}+1\right) \ge \alpha$$

$$\Rightarrow (2d-r-c) \ge 2\alpha(d-c)$$

$$\Rightarrow 2d(1-\alpha) + c(2\alpha-1) \ge r$$

Here the maximum possible value of $\text{Tr}\{\xi \ge r\} \ge \alpha$ will be $\dfrac{1}{2}\left(\dfrac{d-a}{d-c}+1\right)$, i.e., $\dfrac{2d-a-c}{2(d-c)}$ and the minimum value will be 1. So, the value of α must be greater than equal to $\dfrac{2d-a-c}{2(d-c)}$.

Case 3: For $a \le r \le b$, from definition (9)

$$\text{Tr}\{\xi \ge r\} \ge \alpha \;\Rightarrow\; \frac{1}{2}\left(\frac{d-r}{d-c}+\frac{b-r}{b-a}\right) \ge \alpha$$

$$\Rightarrow \frac{(d-r)(b-a)+(d-c)(b-r)}{(d-c)(b-a)} \ge 2\alpha$$

$$\Rightarrow d(b-a)+b(d-c)-r\big[(b-a)+(d-c)\big]$$
$$\ge 2\alpha(d-c)(b-a)$$

$$\Rightarrow \frac{d(b-a)+b(d-c)-2\alpha(d-c)(b-a)}{(b-a)+(d-c)} \ge r$$

Here the maximum possible value of $\text{Tr}\{\xi \ge r\} \ge \alpha$ will be $\dfrac{d-b}{2(d-c)}$ and the minimum value will be $\dfrac{2d-a-c}{2(d-c)}$.

Theorem 3:

If $\xi = ([a,b],[c,d])$ be a rough variable with $c \le a < b \le d$, then for any predetermined α, $\text{Tr}\{\xi \le r\} \ge \alpha$, for $0 < \alpha \le 1$ is equivalent to

(i) $(1-2\alpha)c + 2\alpha d \le r,$ if $\alpha \le \dfrac{a-c}{2(d-c)};$

(ii) $2(1-\alpha)c + (2\alpha-1)d \le r,$ if $\alpha \ge \dfrac{b+d-2c}{2(d-c)};$

(iii) $\dfrac{c(b-a)+a(d-c)+2\alpha(b-a)(d-c)}{(b-a)+(d-c)} \le r,$ otherwise.

The proof is similar to Theorem 2.

Using the above theorems, the crisp equivalent of the model (2) is

$$\begin{cases}
\text{Minimize} \quad \left[\displaystyle\sum_{i=1}^{m}\sum_{j=1}^{n} x_{ij} \mathrm{E}(\tilde{c}_{ij})\right] \\[2mm]
\text{Subject to} \\[1mm]
\quad \displaystyle\sum_{j=1}^{n} x_{ij} \le \mathrm{P}_{\alpha_i}, \; i = 1,2,\dots,m \\[3mm]
\quad \displaystyle\sum_{i=1}^{m} x_{ij} \ge \mathrm{P}_{\beta_j}, \; j = 1,2,\dots,n \\[2mm]
\quad x_{ij} \ge 0, \quad i = 1,2,\dots,m \;,\; j = 1,2,\dots,n
\end{cases} \qquad (11.3)$$

where

$$P_{\alpha_i} = \begin{cases} (1-2\alpha_i)d_i + 2\alpha_i c_i & \text{when } \alpha_i \leq \dfrac{d_i - b_i}{2(d_i - c_i)} \\[3mm] 2(1-\alpha_i)d_i + (2\alpha_i - 1)c_i & \text{when } \alpha_i \geq \dfrac{2d_i - a_i - c_i}{2(d_i - c_i)} \\[3mm] \dfrac{d_i(b_i - a_i) + b_i(d_i - c_i) - 2\alpha_i(b_i - a_i)(d_i - c_i)}{(b_i - a_i) + (d_i - c_i)} & \text{otherwise} \end{cases}$$

$$P_{\beta_j} = \begin{cases} (1-2\beta_j)c_j^1 + 2\beta_j d_j^1 & \text{when } \beta_j \leq \dfrac{a_j^1 - c_j^1}{2(d_j^1 - c_j^1)} \\[3mm] 2(1-\beta_j)c_j^1 + (2\beta_j - 1)d_j^1 & \text{when } \beta_j \geq \dfrac{b_j^1 + d_j^1 - 2c_j^1}{2(d_j^1 - c_j^1)} \\[3mm] \dfrac{c_j^1(b_j^1 - a_j^1) + a_j^1(d_j^1 - c_j^1) + 2\beta_j(b_j^1 - a_j^1)(d_j^1 - c_j^1)}{(b_j^1 - a_j^1) + (d_j^1 - c_j^1)} & \text{otherwise} \end{cases}$$

Model (3) becomes a standard deterministic transportation model that Simplex can solve.

11.4 NUMERICAL EXAMPLE

Consider a problem with three sources $A_i (i = 1,2,3)$ and three destinations $B_j (i = 1,2,3)$. The unit transportation cost, the supply at each source, demands of each destination are rough variables.

Supply ζ_i at the source a_i and demand η_j at the destination B_j are rough variables which are given as follows:

$$\xi_1 = ([20,22][19,23]), \ \xi_2 = ([17,18]\ [16,19]), \ \xi_3 = ([24,25][23,27])$$

Table 11.1 Unit Cost Matrix of Transportation Table With Rough Values

	B_1	B_2	B_3
A_1	([9, 12] [5, 15])	([22, 25] [18, 28])	([33, 35] [28, 40])
A_2	([28, 32] [25, 35])	([28, 32] [25, 35])	([12, 15] [9, 16])
A_3	([38, 42] [35, 45])	([22, 25] [18, 28])	([33, 35] [28, 40])

$$\eta_1 = ([11,13][10,14]),\ \eta_2 = ([24,26]\ [23,27]),\ \eta_3 = ([19,20][18,21])$$

Taking $\alpha_i = 0.9$, $\beta_j = 0.9$, where $i = 1, 2, 3$ and $j = 1, 2, 3$.

Then, model (3) is equivalent to the following mathematical model:

$$
\begin{cases}
\text{Minimize} & \left[\displaystyle\sum_{i=1}^{3}\sum_{j=1}^{3} x_{ij}\,\mathrm{E}\big(\tilde{c}_{ij}\big)\right] \\[2ex]
\text{Subject to} & \\[1ex]
& \displaystyle\sum_{j=1}^{3} x_{ij} \le P_{\alpha_i},\quad i = 1,2,3. \\[2ex]
& \displaystyle\sum_{i=1}^{3} x_{ij} \ge P_{\beta_j},\quad j = 1,2,3. \\[2ex]
& x_{ij} \ge 0,\qquad i = 1,2,3.\ ,\ j = 1,2,3.
\end{cases}
$$

Using the simplex method, the optimal solution is 1089.45, and the corresponding transportation plan is

$$x_{11} = 13.2, x_{12} = 2.4, x_{13} = 3.8, x_{23} = 16.6, x_{32} = 23.8.$$

11.5 CONCLUSION

In the absence of sufficient data to estimate the probability distribution of the random variables, the subjective estimations of the parameters are made by domain experts. Likely, the subjective estimations are usually made in a certain range of values which rough variables of the form can characterize $([a,b],[c,d])$ A rough variable can also characterize an interval estimate $[a,b]$. In this work, the transportation model is developed by considering the unit cost of transportation, supplies, and demands as rough variables to deal with human uncertainty effectively. Furthermore, in this work, expected value of the objective function is minimized; some other form of the objective function can be taken in future work.

REFERENCES

Chanas S., Kolodziejczyk W., and Machaj A., A fuzzy approach to the transportation problem, *Fuzzy Sets and Systems*, 139 (1984), 211–221.

Dantzig G., *Application of the simplex method to a transportation problem, in activity analysis of production on allocation* (1951), John Wiley and Sons, New York, 359–373.

Guo H., Wang X., and Zhou S., A transportation problem with uncertain costs and random supplies, *International Journal of e-Navigation and Maritime Economy*, 2 (2015), 1–11.

Hitchcock F., The distribution of a product from several sources to numerous locations, *Journal of Mathematics and Physics*, 20 (1941), 224–230.

Koopmans T., *Optimum utilization of the transportation system*, Proceeding of the International Statistical Conference (1947), Washington.

Kundu P., Kar M.B., Kar S., Pal T., and Maiti M., A solid transportation model with product blending and parameters as rough variables, *Soft Computing* (2015), 1–10. DOI:10.1007/s00500-015-1941-9.

Liu B., Inequalities and convergence concepts of fuzzy and rough variables, *Fuzzy Optimization and Decision Making* 2 (2003), 87–100.

Liu B., *Uncertainty theory: An introduction to its axiomatic foundation* (2004), Springer-Verlag, Berlin, Germany.

Liu B., *Uncertainty theory* (2007), Springer-Verlag, Berlin, Germany.

Liu B., Some research problem in uncertainty theory, *Journal of Uncertainty Systems* 3 (2009a), 3–10.

Liu B., *Theory and practice of uncertainty programming*, 2nd ed. (2009b), Springer-Verlag, Berlin, Germany.

Liu B., Uncertain set theory and uncertain inference rule with application to uncertain control, *Journal of Uncertainty Systems*, 4 (2010a), 83–98.

Liu B., *Uncertainty theory: A branch of mathematics for modeling human uncertainty* (2010b), Springer-Verlag, Heidelberg, Germany.

Pawlak Z., Rough sets, *International Journal of Information and Computer Science*, 11(5) (1982), 341–356.

Sheng Y. and Yao K., A transportation model with uncertain costs and demands, *An International Interdisciplinary Journal*, 15(8) (2012a), 3179–3186.

Sheng Y. and Yao K., Fixed charge transportation problem and its uncertain programming model, *Industrial Engineering and Management Systems*, 11(2) (2012b), 183–187.

Subhakanta D., and Mohanty S. P., Uncertain transportation model with rough unit cost, demand and supply, *OPSEARCH* (2019), 1–13. DOI:10.1007/s12597-017-0317-6.

Yang L. and Liu L., Fuzzy fixed charge solid transportation problem and algorithm, *Applied Soft Computing*, 7 (2007), 879–889.

Williams A.C., A stochastic transportation problem, *Operations Research*, 11 (1963), 759–770.

Chapter 12

Impact of COVID-19 on the world economy

Altaf Ahmad, Anjanna Matta, and Topunuru Kaladhar

12.1 INTRODUCTION

The COVID-19 episode was initiated in December 2019 in the city of Wuhan, which is in the Hubei region of China. The infection keeps on spreading over the world. The point of contact of the eruption was at first China, with announced cases either in China or in explorers from the nation; after that, the cases were accounted for in numerous other nations. While a few countries have had the option to treat announced cases adequately, it is unsure where and when new cases will rise. Despite the enormous open well-being hazard COVID-19 stances to the world, the World Health Organization (WHO) has pronounced a global well-being crisis of global worry to organize worldwide reactions to the disorder.

Nonetheless, it is discussed whether COVID-19 could conceivably heighten a worldwide pandemic. In a strongly associated and incorporated world, the effects of the ailment's past mortality (the individuals who kick the bucket) and horribleness (those who can't work for a time) have become obvious since the episode. In the midst of the easing back down of the Chinese economy with interference to creation, the working of worldwide gracefully chains has been disturbed [1]. Organizations worldwide, independent of the size that are reliant upon contributions from China, have begun encountering constrictions. Transport being constrained and confined among nations has also eased worldwide financial exercises. Specifically, some free-for-all among buyers and firms has ruined typical usage structures and created market inconsistencies. Overall budgetary markets have also been responsive to the movements, and stock records have plunged. The coronavirus infection 2019 is delivered by another disease for which, right now, there is no pharmaceutical treatment. The elements of the malady are with the end goal that, without non-pharmaceutical mediations (NPIs), it overpowers the limit of national social insurance frameworks. Thus, governments decided to establish NPIs to contain the spread of the COVID-19 pandemic.

We initially depict the central issues that developing economies at present face. It considers the direct and circuitous monetary expenses of NPIs. We contend that the immediate cost of NPIs could be critical: over 20% of GDP over the period in which NPIs are set up. If NPIs continue, these immediate expenses are exacerbated by roundabout costs: numerous family units and firms must keep paying fixed expenses while their salaries fall [2]. This passing fall in pay, combined with the vulnerability about how long the pay stun will last, will prompt a huge increment in the interest for liquidity. We are, as of now, observing sizable portfolio shifts in USA resource markets. The budgetary pressure caused by the steady confound among pay and costs will probably bring about an expansion in joblessness, charge delays, and obligation restructurings. For rising economies, NPIs to contain the spread of COVID-19 are being presented while ware costs are falling (25% up until now) and sovereign credit spreads are expanding. Developing economies running record shortfalls will probably encounter an unexpected stop in capital streams. This stun is known to cause extreme downturns in rising economies. The strategy suggestions are that current NPIs are monetarily impractical. Interest in more effective manners of distinguishing people requiring disconnection is basic [3]. This is a most worry for the entire world, the mission for more proficiency and focused on NPIs thought to be a worldwide, multilateral, and helpful undertaking. We finish up with remarks on the monetary strategy ramifications of the current circumstance. We underline the strife between the financial pressure looked at by open accounts in developing economies and the requirement for guaranteed palliative economic approaches. Governments face lost income, an expansion in the interest for unrestricted use, also fixed worldwide money-related conditions. The industriousness of NPIs is a budgetary delayed bomb for the private area and concerning sovereigns [4].

In considering the strategic responses to the test introduced by this overall gathering of stars of shocks, we encourage governments to assess how much use they can manage the expense of emergency spending bills and to recall for them a future fiscal change for when the plague is done. This arranging fuse as expected wellsprings of advantages propels from genuine multilateral crediting establishments and the money-related influence. Countries with a tremendous current commitment organization weight may consider sovereign commitment restructurings [5]. We believe the neighborhood courses of action are stretched out development undertakings to firms and nuclear families in the best possible region and moves to administrators in the easygoing part. This is the first of plenty of notes to be released by UNDP. Neither country express methodologies nor the distributive impact of COVID-19 is thought of. As the improvement office of the United Nations, UNDP has a long custom of going with policy-making in its structure, execution, checking, and assessment. It has an order to react to evolving conditions, conveying its resources to help our part states in their quest for coordinated answers to complex issues [6]. This arrangement targets drawing from UNDPs own experience and information internationally and from the mastery, furthermore, the limit of our accomplice thinks tanks and scholarly organizations in Latin America and the Caribbean. It is an endeavor to advance an aggregate reflection on the reaction to the COVID-19 well-being emergency and its monetary furthermore social consequences on our social order. Idealness is an absolute necessity. Arrangements that depend on proof, understanding, and contemplated strategy instinct—originating from our rich history of strategy commitment are essential to direct this exertion. This arrangement adds to the incorporated methodology set up by the UNO change. It contributes significantly to the conscious reaction of the United Countries' improvement framework at the worldwide, local, and national levels. Interestingly, the compelled predicament on interior development and better economic strategy spending affected the diploma of monetary and physical games, despite the reality that the increasingly wide variety of asserted coronavirus instances failed to affect the certification of financial sports drastically.

12.1.1 The primary financial issues looked at by rising economies

The NPIs intended to contain the COVID-19 pandemic, similar to limitations on the development of individuals and social removal measures, are required to affect worldwide financial action significantly [7]. This worldwide stun has general balance consequences for costs that affect business cycles in rising economies. COVID-19 influences developing economies through three virtual channels:

I. The direct impact of NPIs on monetary movement due to:
- Limitations to the yield of numerous enterprises include travel and amusement.
- Limitations on social contact constrain a few people to telecommute or not work by any stretch of the imagination, bringing down the yield.

II. **Terms of exchange:** numerous item trading nations are encountering a sharp fall in the costs of the products they send out, influencing a sizable division of GDP and government incomes.

III. **Worldwide money-related stun:** A global liquidity stun involves monstrous portfolio shifts from less secure to more secure fluid resources. For developing economies, this suggests capital outpourings, an expansion in their financing expenses, and a drop in the estimation of their monetary forms.

12.1.2 The direct impact of NPIs

It is too soon to precisely gauge how much the immediate expense of NPIs will be. We can guess. Starter information and back-of-the-envelope estimations show that they could be significant. In China, the drop in mechanical creation between December 2019 and February 2020 was nearly 25%. Investment in fixed resources, a check of development movement, slid 24.5% during a similar period, turning around the development of 5.4% in 2019. Retail deals tumbled 20.5% in the initial two months of the year—ordinarily a blast season for utilization contrasted with and development of 8.0% in December 2019 [8].

As of 21 March 2020, more than 271,364 coronavirus malady (COVID-19) have been affirmed across 174 nations and regions. Continued human-to-human transmission has now been seen in countries outside China, including Italy, Japan, and South Korea, with 47,021, 1,007, and 8,799 cases revealed, respectively [9]. Then again, a few nations, for example, Bangladesh, have all later announced their first instances of COVID-19 coming about because of importations of contaminated explorers from influenced territories. Accordingly, nations and districts have executed a broad scope of non-pharmaceutical intercessions (NPIs).

We are anxious that these figures might have troubling drawbacks and dangers. A few back-of-the-envelope counts lead to this end. (a) Italy, Spain, Argentina, New York, Illinois, and California have been wholly secured for (up until this point) 14 days. An exceptionally ideological situation is that monetary action is down to half what is typical for about 14 days, and afterward, it promptly bounces back [10]. This would suggest a quarter-on-quarter fall of 8.3% with a lockdown as for a typical quarter. Three of the lockdown is for three weeks rather than two; the quarter-on-quarter fall ascends 12.5%. (b) In a time of social separation, 40% of the economy works at half typically. The inferred drop in total yield is 20%. Table 12.1

Table 12.1 Hypothetical Percentage Change in GDP [11]

Output drop in affected sector	Share of output affected by NPIs (%)								
	10%	20%	30%	40%	50%	60%	70%	80%	90%
10%	-1%	-2%	-3%	-4%	-5%	-6%	-7%	-8%	-9%
20%	-2%	-4%	-6%	-8%	-10%	-12%	-14%	-16%	-18%
30%	-3%	-6%	-9%	-12%	-15%	-18%	-21%	-24%	-27%
40%	-4%	-8%	-12%	-16%	-20%	-24%	-28%	-32%	-36%
50%	-5%	-10%	-15%	-20%	-25%	-30%	-35%	-40%	-45%
60%	-6%	-12%	-18%	-24%	-30%	-36%	-42%	-48%	-54%
70%	-7%	-14%	-21%	-28%	-35%	-42%	-49%	-56%	-63%
80%	-8%	-16%	-24%	-32%	-40%	-48%	-56%	-64%	-72%
90%	-9%	-18%	-27%	-36%	-45%	-54%	-63%	-72%	-81%

shows how NPIs influence various blends of the weight of areas and their yield drops in real influence GDP [11]. A more exact estimation of the effect of COVID-19 on GDP is to gauge its immediate impact on every specific part and afterward to utilize input-yield grids to follow the reaction to the stuns on different economic divisions. (c) With a work share in yield of about ⅔, if successful work hours fall by 30%, the yield will drop about 20%. These numbers are overwhelming. Expanded times of wide social removal measures could be amazingly exorbitant.

This table shows the rate of change in GDP. It is a component of the portion of the influenced part in GDP and the size of the breakdown in those parts.

12.1.3 Indirect impact of NPIs

The immediate effect of NPIs in the past area doesn't consider the potential second-round impacts of this stun. There will probably be a progression of complex to-measure circuitous consequences, further diminishing monetary movement. In a social removal condition, numerous organizations experience negative worth, including as the expense of data sources surpasses net creation. Firms can't sell their products and enterprises; however, they, despite everything, need to take care of the pay tab, administration their obligations, and make good on rent and expenses [12]. Broadened times of NPIs will have a few further dangerous consequences for the economy.

1. Numerous organizations leave business: This is particularly valid for firms in serious social contact ventures (travel and diversion) and small and medium undertakings (SMEs) with small working capital and restricted credit lines. Restarting these organizations might be a lengthy and exorbitant procedure.
2. Firms exhausting their capital will cut back laborers: We know from past downturns that after spikes in joblessness, coordinating laborers and opening in the recovery are a moderate procedure.
3. Cutbacks decrease total interest.
4. Most of Families and firms confronting expanded vulnerability spare more in the most secure resources. They also entered dangerous zones socially and economically.
5. Limitations on monetary action and the cutoff points on the development of individuals reshape flexible chains and creation systems with lost productivity.
6. New types of (working from home) may likewise decrease proficiency. In this situation, many developing countries' economies suddenly fall. Which created two different stocks: one is proficiency, and the other is economics.

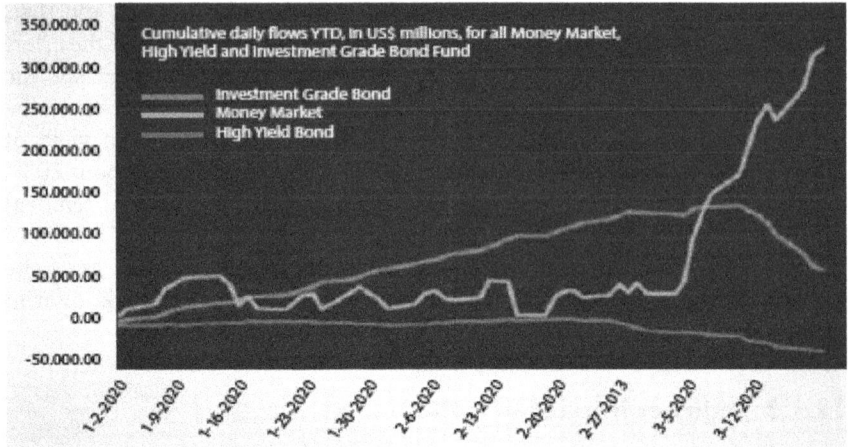

Cumulative daily flows YTD, in US$ millions, for all Money Market, High Yield and Investment Grade Bond Fund

Investment Grade Bond
Money Market
High Yield Bond

Figure 12.1 Sudden increase in the demand for liquidity (USA) [12].

Figure 12.1 shows how, in the USA, private operators are hauling cash out of credit markets into currency to advertise shared reserves that the Fed truly backstops. We anticipate a comparative increment in the interest for most fluid resources in Latin America.

12.1.4 Commodity prices

For commodity exporters, this stun alone would be trailed by a sharp cash devaluation and a downturn. For nations where product sends out are a significant wellspring of government income, for instance, Argentina, Bolivia, Chile, Colombia, Ecuador, and Mexico, this stun to the terms of exchange will likewise strain open accounts [13]. Commodity shippers in Latin America and the Caribbean will profit from this stun.

Figure 12.2 shows a record of commodity costs (oil, soybean, copper, espresso, and so forth). Between January and February, item costs have fallen by around 25%.

12.1.5 Existing influence represents an extra hazard

After close to zero financing costs, partnerships and sovereigns are stacked with the obligation. If NPIs persevere and these foundations can't administration or rollover their obligation, a worldwide money-related emergency associated with or more regrettable than the one of 2008–9 could occur [14]. Overleveraged sovereigns hit by the monetary results of the downturn and the product value drop in a world with tight budgetary conditions may be enticed to rebuild their obligations. The expectation of these choices may transform

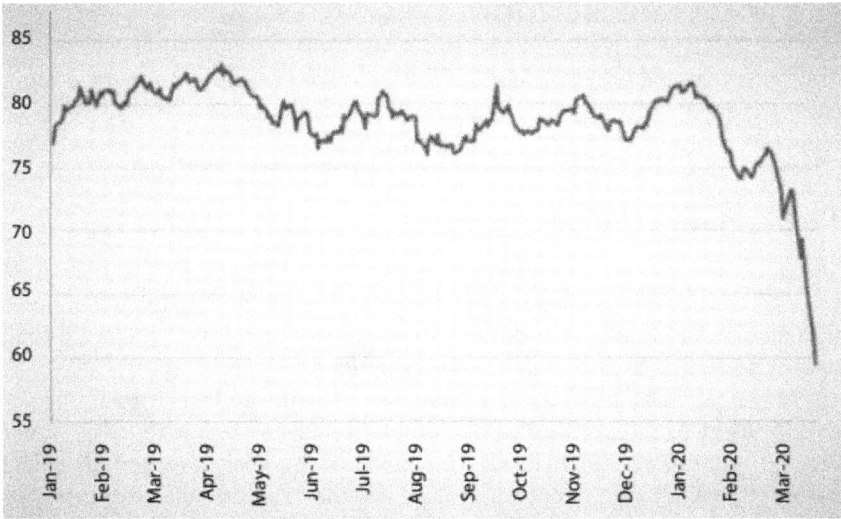

Figure 12.2 Index of commodity prices [13].

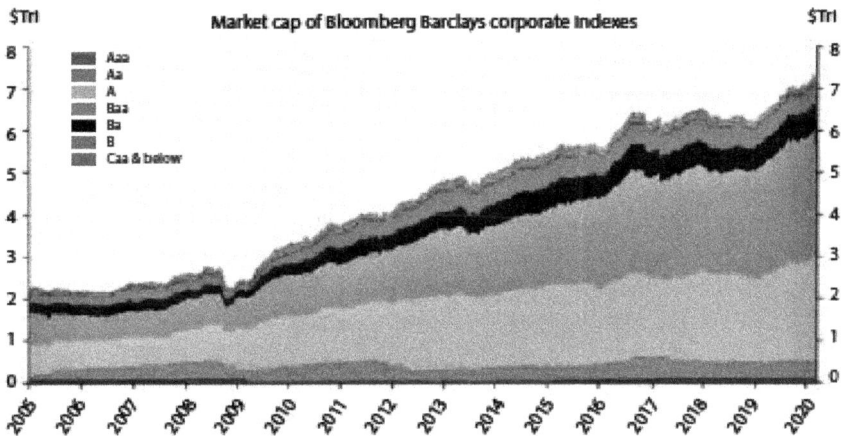

Figure 12.3 Corporate dept.

into an unavoidable outcome. In this setting, it would be essential to evaluate worldwide open parts getting necessities against the IMF's capability.

Figure 12.3 delineates the expansion of nonfinancial corporate obligations in the USA. On the off chance of a tenacious downturn, B-evaluated obligation will probably be minimized. Transient corporate obligation (31% of all out obligations) remains at about 10% of GDP [15]. Family obligation administration in the USA is on request for 10% of individual extra cash. The administration of these obligations probably won't be turned over.

12.2 POLICY IMPLICATION

The NPIs that are presently executed in numerous nations have huge monetary expenses. This segment examines diverse approach reactions for governments in developing economies' markets.

12.2.1 More gainful NPIs

Social orders are receiving extraordinary NPIs to contain the spread of COVID-19, because epidemiological models foresee that medicinal services frameworks intended for ordinary occasions can't deal with the additional weight presented by intense COVID-19 patients.

These epidemiological models anticipate that the conversational pace of development of the number of contamination cases is relative to the number $R0_s -1$, where $R0$ is the actual generation number (additionally called essential proliferation proportion), and s is the extent of the populace that is vulnerable (not insusceptible) to the sickness. The proportion $R0$ is the standard number of new contaminations from a solitary tainted person when all of the populace is vulnerable [16]. The item $R0_s$ is the standard number of new contaminations from a solitary person when a portion s of the populace is helpless. It tends to be deciphered as the proportion between the standard recuperation time of an irresistible individual and the normal time between new cases. The scourge develops when the recuperation time surpasses the time between new cases. As we don't know the real number of irresistible cases, just those tried, assessments of $R0$ and of s are uncertain. Likewise, $R0$ is certifiably not an organic steady: it results from an intricate social harmony controlled by the recurrence with which irresistible individuals reach others for a sufficiently long time to communicate the infection. $R0$ relies upon arrangements and accepted practices and can change across nations, districts, and times inside a locale [17]. NPIs, for example, the stay-at-home strategies in Italy, Spain, Argentina, California, Illinois, and New York, are extraordinary measures to decrease contact rates among individuals and, consequently, the fundamental conceptive number $R0$.

- The best approach to mediation to contain COVID-19 is to build up innovation to lessen the contact rate among irresistible and helpless people while confining society-wide human collaboration to as meager as expected. Given the enormous expenses of widespread lockdowns, the pace of profit for speculations to empower focused on arrangements of disconnection is considerable [18]. Directed NPIs would detach a subset of people (for instance, irresistible people, people that are probably going to be irresistible, and the weaker defenseless).
- Two activities toward this path are forceful trying to distinguish between irresistible and insusceptible people. This information will permit invulnerable people to flow openly and work. Destructive

testing additionally assists with contrasting irresistible people early and following their contacts. Hong Kong, Iceland, Japan, Singapore, and South Korea executed directed detachment strategies.

- Collecting a group of analysts and disease transmission experts to configure testing procedures and now cast the actual conceptive number across geological and social bunches could assist governments with allocating clinical assets and social intercession strategies all the more effectively [19, 20, 21, 22]. This information will help with structuring proof-based NPIs.
- Forceful testing could likewise empower strategy creators to assess (progressively) the adequacy of various social separating intercessions (shutting schools, shopping centers, games, etc.). Even though general NPIs are very expensive, there is no adequate assessment of various intercessions.

12.3 CONCLUSION

In this book chapter, complete analyses of money flow, environmental issues, and socio-economic impacts in the context of fitness emergency activities and the worldwide responses to mitigate the results of these occasions were furnished. COVID-19 is a global pandemic in that locations save you from financial hobbies and pose a severe danger to overall well-being. The global socio-monetary impact of COVID-19 consists of higher unemployment and poverty fees, decreased oil expenses, altered schooling sectors, modifications within the nature of work, decreased GDPs, and heightened dangers to healthcare people. Thus social preparedness, as a collaboration among leaders, fitness care personnel, and researchers to foster extensive partnerships and devise strategies to gain socio-economic prosperity, is needed to address future pandemic-like conditions. The impact at the power location consists of extended residential energy name for a reduction in mobility and alternative inside the nature of labor. Lockdowns throughout the globe have confined movement and characteristics located at home, which has, in turn, decreased commercial enterprise and industrial energy calls for similar waste technology. Sustainable metropolis management that highlights the outstanding advantages of ecological stability is essential to lower viral infections and exceptional sicknesses. Policies that sell sustainable improvement ensure that cities can be implemented and encourage measures like social distancing and self-isolation will deliver a mean benefit immediately. The first era of COVID-19 vaccines is predicted to gain approval with the resource of the surrender of 2020 or early 2021, so you can provide immunity to the populace. It is essential to set up preventive epidemiological fashions to discover the prevalence of viruses like COVID-19 earlier.

Moreover, states, policymakers, and partners around the area need to do whatever it takes to defeat this Coronavirus circumstance, alongside ensuring medical care administrations for all occupants, helping people who are likely working in cutting-edge administrations and experiencing huge money-related impacts, ensuring social removing, and focusing on developing a feasible future.

REFERENCES

[1] Igor Rudan (2020) A cascade of causes that led to the Covid-19 tragedy in Italy and in other European union countries. *Journal of Global Health*, 10(1).

[2] Ben Wedeman (2002) *Covid-19 has taken a staggering toll on a whole generation in Northern Italy* www.cnn.com/2020/05/21/europe/italy-nursing-homes-deaths-intl/index.html

[3] Tim Lister and Claudia Rebaza (2020) *How Spain became a hotspot for coronavirus* www.cnn.com/2020/03/28/europe/spain-coronavirus-hotspot-intl/index.html

[4] Coronavirus: Brazil headed for catastrophe (2020) www.dw.com/en/coronavirus-brazil-headed-for-catastrophe/a-53502907

[5] Maria Nicola, Zaid Alsafi, Catrin Sohrabi, Ahmed Kerwan, Ahmed Al-Jabir, Christos Iosifidis, Maliha Agha and Riaz Agha. (2020) The socio-economic implications of the coronavirus and Covid-19 pandemic: A review. *International Journal of Surgery*, 78: 185–193.

[6] Scott R Baker, Nicholas Bloom, Steven J Davis, and Stephen J Terry. (2020) *Covid-induced economic uncertainty.* Technical Report, National Bureau of Economic Research.

[7] McKibbin Warwick and Peter Wilcoxen (1999) The theoretical and empirical structure of the G-cubed model. *Economic Modelling*, 16(1): 123–148.

[8] Mckibbin Warwick and R Fernando (2020) *The global macroeconomic impacts of COVID-19: Seven scenarios.* CAMA Working Paper Series.

[9] Joanna Wilson (2020) The economic impact of coronavirus: analysis from imperial experts. *Imperial News.*

[10] Warwick Mckibbin and David Levine (2020) Simple steps to reduce the odds of a global Catastrophe. *Australian Financial Review.*

[11] Nuno Fernandes (2020) *Economic effects of coronavirus outbreak (covid-19) on the world economy.* Available at SSRN 3557504. IESE Business School Working Paper No. WP-1240-E.

[12] Edward Lempinen (2020) www.economic-impact-human-solutions

[13] Coronavirus: Travel restrictions, border shutdowns by country www.aljazeera.com/news/2020/03/coronavirus-travel-restrictions-border-shutdowns-country-200318091505922.html

[14] Coronavirus declared a pandemic as fears of economic crisis mount (2020) www.ft.com/content/d72f1e54-6396-11ea-b3f3-fe4680ea68b5

[15] Coronavirus: Sport-by-sport look at the global impact of COVID (2020) www.independent.co.uk/sport/sport-football-basketball-rugby-olympics-cancelled-coronavirus-impact-around-the-world-a9398186.html

[16] T Buck, M Arnold, G Chazan, C Cookson (2020) *Coronavirus declared a pandemic as fears of economic crisis mount* www.ft.com/content/d72f1e54-6396-11ea-b3f3-fe4680ea68b5

[17] Jennifer Meierhans (2020) *Coronavirus: What are independent supermarkets doing to help?* www.bbc.co.uk/news/uk-england-51947391

[18] Coronavirus (COVID-19) (2020) Harvard University www.harvard.edu/coronavirus

[19] Warwick McKibbin and Roshen Fernando (2020) *The global macroeconomic impacts of Covid-19*. Brookings Institute, pages 1–43.

[20] A Khoo (2020) *Coronavirus lockdown sees air pollution plummet across UK* https://www.bbc.com/news/uk-england-52202974

[21] R Lillywhite (2020) *Air quality and well-being during COVID-19 lockdown* https://www.newswise.com/coronavirus/air-quality-and-wellbeing-during-covid-19-lockdown/?Article_id=730455

[22] J Brunton (2020) *Nothing less than a catastrophe': Venice left high and dry by coronavirus* https://www.theguardian.com/travel/2020/mar/17/nothing-less-than-a-catastrophe-venice-left-high-and-dry-by-coronavirus

Chapter 13

Food waste

Impact of COVID-19 on urban and rural areas

*Surya Kant Pal, Mahesh Kumar Jayaswal,
Subhodeep Mukherjee, Kriss Gunjan, Jyotirmai
Satapathy, and Simran Singh*

13.1 INTRODUCTION

Infectious diseases burden communities and societies throughout the world. As an infectious disease increases in any population, people begin to look for methods most effective in combating the outbreak or at least controlling the number of infections (Lu et al., 2022). Scientists have made tremendous progress in the fight against diseases. Yet infectious diseases remain a significant cause of mortality. In epidemiology, one aims to investigate improving well-being and conditions in a specific population to control related-health problems. This thesis uses mathematics to describe complex disease dynamics using simplifications and hypotheses about the relevant mechanisms (Filimonau & Uddin, 2021). The pandemic's progression also has brought up the issue of food security. Although no significant food scarcity was reported, some disruptions started due to limitations on people's movement, the closure of certain businesses, restrictions on imports and exports, and the impairment of freight transportation. The pandemic affects all four pillars of food security: accessibility, access, consumption, and consistency.

Regarding this first stage, there were premises to reduce food waste: decreased income in many families due to lockdown unemployment; price increases in only certain foods; and frequent family cooking due to more time spent at home and restaurants closing (Fan et al., 2022). The net effect of the pandemic on food waste will be ascertained by its longevity, the impact on the world economy, the agri-food supply chain, and households, as well as local government measures and regional, national, and global pandemic management. Food loss seriously affects food security, the surroundings, and the global, regional, and efficient financial system (Kharola et al., 2022).

The spread of the corona pandemic, along with human health and wealth, also affected the food sector. There is a continuous increase in the infection of COVID-19 globally through various reports of food workers all over the world getting infected. Many restaurants and retailers have started focusing on curbside pickups (Ciccullo et al., 2022). According to Adobe Analytics, buy-online-pickup-in-store orders have increased 87% YoY between late February and 29 March. Quick service chains like McDonald's are improving their takeout facility and drive-through services, decreasing their waiting time and friction to deliver customers a better experience (Ciccullo et al., 2022).

So, while the world is seeing business owners in the food industry go through economic challenges, we're also witnessing great examples of new initiatives, community-building efforts, and a collective and empathetic stand to fight COVID-19 together. Due to the COVID-19 crisis, every industry in the urban areas expects to see how this pandemic outbreak will affect the manufacturing industry (Costa et al., 2022). A certain number of people working in the manufacturing and food industry have the potential to starve, but if any of the staff members are infected, all the people are at risk (Filimonau et al., 2022). The rapid spread of the infections aroused various myths like transmission of the virus can occur from the touch of person to person (Baral, Mukherjee, Nagariya, et al., 2022; Mukherjee, Baral, Chittipaka, et al., 2022).

Our study will examine the gap not included in food waste in urban and rural areas of India. Its purpose is to make people aware of the food waste in the urban and rural areas, especially in the essence of COVID-19, the impact of COVID-19 on the urban and rural areas were very different, or it should be addressed as opposite, as the metropolitan area started to use the online platform to order foods and vegetables, on the other hand, some rural area very thriving for food as they have no money to buy. If I compare previews two years, according to Global Hunger Index (GHI, 2020), India ranked 94th out of 107 countries, whereas 101st out of 116 countries in 2021 (GHI 21). According to the food waste index report 2021 by the "United Nations environment program (UNEP)", India's waste was recorded to be 90 kg per capita per year.

- Our objective is to make people more aware of food waste and help them understand what to do with the leftover food (in urban and rural areas).
- The importance of reducing food waste for improving sustainable development goals (SDGs) by the end of 2030.
- Our research includes young and elder minds of urban and rural areas and what were there points the technologies used during and after COVID-19 by food industries were helpful for them, making them aware of government schemes like PM Kisan SAMPADA Yojana, which are for storage of mass production for farmers.
- People mostly stocked food items in urban areas, leading them to unplanned shopping and buying extra that they don't need. In contrast, in rural areas, people use their food with very little, so they don't waste it. If the pandemic hits again, they know how to face it.

- COVID-19 may be a wake-up call for people to get aware of food wastage. We told them that if they have leftovers, it's better to use a fridge to stock them for longer or give them to their pets.
- According to Srishti Jain, co-founder of Feeding India, they take donated foods and provide them to the center to ensure that the food reaches the needy.

13.2 LITERATURE REVIEW

13.2.1 Food waste behavior in the household

Food waste is a significant factor contributing to a major threat to people and the environment, possibly requiring well-organized disposal systems. Food and nutrition security challenges can be resolved by food waste reduction, conserving the environment, and providing healthy foods (Yan et al., 2022). It was noticed that roughly a million tons of food are going to waste. People are aware of the issues associated with sustainable food waste, but unintended procedures contribute to excessive food waste. Food waste in homes, considered greater than in other businesses, tends to range from 28.4% to 31.9%. Consumers, irrespective of whether they are the principal source of food waste, necessitate a complex set of behavioral management patterns (Jenkins et al., 2022). It has been recommended that the youngsters be monitored closely to understand their intentions and behaviors toward food waste in developing education programs to limit their food waste practices (Cakar, 2022). The appearance of a young age group in a household increases food wastage in the home. Youth customers are increasingly concerned about the financial aspects of food waste than the environmental and health impacts (Mukherjee, Chittipaka, Baral, & Srivastava, 2022; Mukherjee, Venkataiah, Baral, et al., 2022). Moreover, because they are unaware of reusing food scraps, younger customers may not maximize their food consumption (Strotmann et al., 2022). Besides that, because young consumers lack experience, they commonly miscalculate meal portion sizes, likely resulting in thrown-away leftovers (Cudjoe et al., 2022).

Food waste in India is a vague problem, but some people ignore it. During COVID-19, people started to waste less food. Instead, they began to stock food because of the lockdown imposed by the government of India, and the stocking of food led to expiring food wastage. The end of the pandemic is difficult to predict, affecting public health, industries, financial markets, and infrastructure.

In specialized literature, food waste is a very under-research topic in India. Most of the research is limited to weddings and restaurants. There is less talk about household food waste before and after COVID-19. Food waste awareness in young people and how they participate directly and indirectly in the food system. This chapter is based on data collected in Romania; the impact of COVID-19 on young people changed their mindset toward food, travel, and health (Deng et al., 2022). Food waste among students is being focused on food waste in the social spaces of university campuses (Borg et al., 2022).

13.3 HYPOTHESIS FORMULATION

The issue of willingness to take responsibility for food waste behavioral patterns, which decrease and prevent food scraps individually, receives little attention in the academic literature. Food waste should be viewed as the result of some behaviors related to a journey of food through into people's homes: food planning, shopping, stockpiling, preparedness, and consumption. Given this, each of these behaviors must be approached as responsible minimize and preventing food waste at the household and individual levels (Durán-Sandoval et al., 2021). They can be directed toward responsible food waste behavior through knowledge, communication, and awareness. Specific food education, acquired in a formal or informal configuration, is a system of principles that may or may not be consistent with the kind of behavior.

People's knowledge and concern about food waste are driven more by economic motivations than environmental and health impacts (Hadjichambis et al., 2022). Public advertisements to reduce food waste must be tailored to different young people groups based on knowledge, awareness, and worry about achieving efficiency. Knowledge of food waste among people is based on awareness, understanding, or information on food waste management that they acquire through experience. One of the main reasons for food waste behavior in urban and rural people is a lack of environmental and financial development awareness (Yousefi et al., 2021). Awareness and concern of youngsters regarding food waste reflect economic development, then ecological concerns (Iranmanesh et al., 2022). Government campaigns should be tailored to the valuable concept of food waste knowledge, concern, and awareness (Mukherjee, Chittipaka, Baral, Pal, et al., 2022). Rural people do not know how to store or prepare different types of food from the same ingredients, whereas urban people know but still make more than they require. Then they tend to throw the leftovers; on the other hand, rural people have little food left over; instead, they give their pieces to animals (Daliakopoulos et al., 2022). Food waste is defined as edible food thrown out from private households. Food waste is seen as a waste of resources, which can be more helpful in the hunger problem that countries suffer. Food waste produces adverse effects regarding limited resources, biodiversity, and climate change (Burlea-Schiopoiu et al., 2021). The pandemic favors food waste reduction: less frequent shopping, well-planned meals, and storing leftovers in fridges. Concerns about healthy and good immunity are determined as people start eating organic products, fruits, and other nutritional products (Yi-Chi Chang et al., 2022).

People become more aware of the extent to which food waste affects the surrounding ecosystems; they constitute the generation that will increase changes and adapt their behavior to the needs of environmental conservation, and it will develop new methods to minimize the negative effect that population habits have on the environment (Diekmann et al., 2021). Irrespective of the factors highlighted in the literature review, changing people's behaviors to reduce food waste and its environmental impact can be accomplished with the assistance of retail outlets and marketing executives. However, customers are still concerned about reducing food waste since they are primarily motivated by financial ramifications in terms of costs and benefits than by environmental protection (Amicarelli & Bux, 2021). The implications of the COVID-19 crisis on food purchases take into account people's healthy buying habits. Shopping for food routines was something of an automatic thing (Obuobi et al., 2022). The crisis compelled a time frame of reflection and changes in behavior that we are attempting to instill in our variable (Mukherjee, Chittipaka, & Baral, 2022). The pandemic ended up causing spontaneous shopping, overstocking, and travel bans, all of which have an impact on shopping routines. Increased consumption of snacks cooked at home at the expense of that consumed city is likely to result in waste because household units lose the ability to measure and standardize portions (Münch et al., 2021). Because connectivity to some products was limited due to overstocking or restrictions on travel during the quarantine period, food stockpiling enhanced, potentially contributing to wasted food by discarding excess inventory (Mukherjee & Chittipaka, 2021).

H1: The influence of the COVID-19 crisis on the food shopping of urban and rural people impacts the behavioral intention for food waste.

H2: The population's mindset toward food waste's environmental effects impacts the behavioral intention for food waste.

H3: Technology for generating food waste in urban and rural areas impacts the behavioral intention for food waste.

H4: Knowledge about food waste among the population impacts the behavioral intention for food waste.

13.4 RESEARCH METHODOLOGY

Samples were collected online as well as offline. We sent the online form through a digital platform invitation to participate in the food waste study. People were requested to fill it out and give us their views on it, which was sent to every age group. Our group then visited (Luharli Greater Noida) for the survey and divided it into subgroups; we interviewed people by people of every age group and collected the data to know their behavior regarding

Table 13.1 Model Summary

Model summary				
Model	R	R square	Adjusted R square	Std. error in the estimate
I	0.652[a]	0.871	0.672	0.21346
2	0.692[b]	0.732	0.712	0.17651
3	0.671[c]	0.765	0.653	0.18973
4	0.712[d]	0.768	0.821	0.13214

food waste in the urban and rural areas (Baral, Mukherjee, Chittipaka, et al., 2022).

We spread awareness of food waste in urban and rural areas by doing surveys online and offline. In online mode, we collected the data from 57 people of every age and gender. During the visit to a village called Luharli, we interviewed villagers of every age. In total, 50 interviews were taken by the group, and we did manage to inform them of food waste management and schemes started by the government of India to store food and crops. We get unique and intriguing answers from them. Our purpose was to tell them how to manage food after COVID-19 is at its lowest.

This research has taken four independent variables and one dependent variable Mukherjee, Baral, Pal, et al., 2022; Roy et al., 2022). Here Table 13.1 shows variables entered or removed. Out of four independent variables, all four significantly contribute to the dependent variable. SPSS 20.0 has been used for the analysis. The variables contributed significantly. All four hypotheses were accepted. Figure 13.1 shows that all the hypothesis are accepted, as R^2 values are more significant than 0.5.

13.5 DISCUSSION

Consumers who feel guilty about throwing away food are more likely to engage in food-saving behavioral patterns such as reducing, reusing, and recycling. People's beliefs and behaviors tend to be shaped once they predict certain emotions. Furthermore, ethical imperatives inform them to alter their behaviors through social learning (Vasseur et al., 2021). Our study suggests that customers are more likely to participate when they know the consequences of reducing food waste. Our results supported the rearrangement of impacts such as economic burden, unnecessary hunger, and climate change, which induce customers to act sustainably and result in food reduction procedures. The COVID-19 pandemic's effect on food waste is combined.

On the one hand, our research identified a connection between the COVID-19 pandemic and reasonable consumption, environmental conservation, and improving the morality of food waste among many young people. On the other side, the insecurity that comes with an emergency can lead

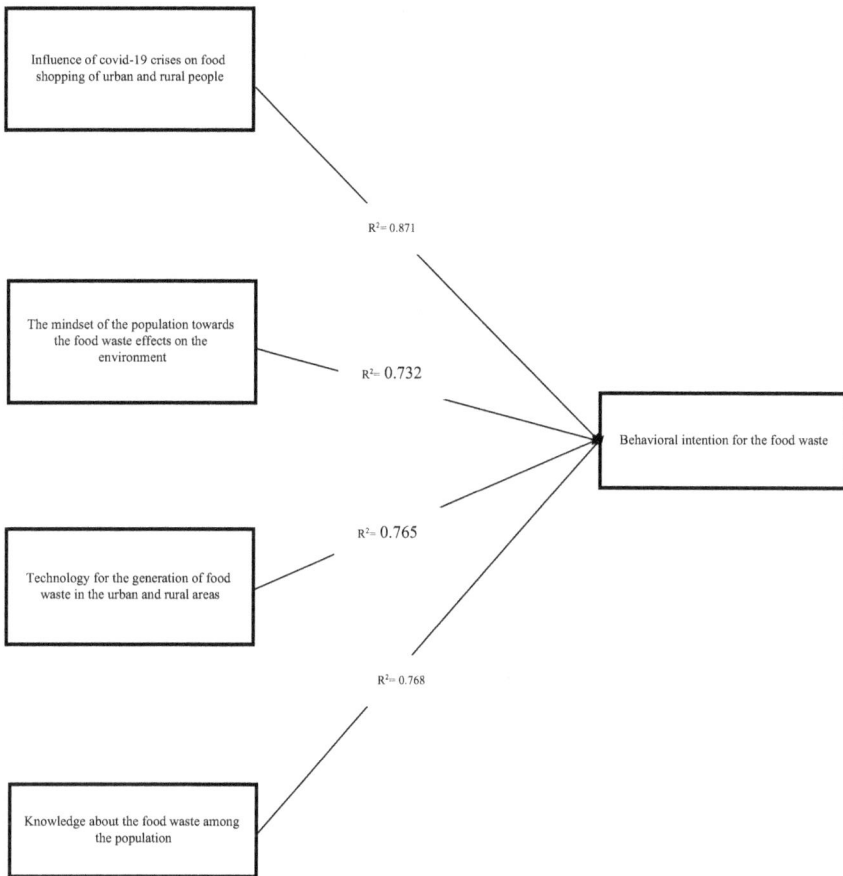

Figure 13.1 Model fit parameters for the behavioral intention for the food waste.

to overconsumption and excess inventory. Consequently, we believe that the adverse effects of the COVID-19 disease outbreak on wasted food were only transitory and that once things returned to "regular", the beneficial benefits recorded in the research will indeed take around (Massari et al., 2022).

For young people, the effect of the COVID-19 disease outbreak on food shopping contributes to re-alignment with individual values, ethical norms, and responsible behavior toward food waste. Food shopping under COVID-19 is a more intentional process requiring needs and a budget (Beheshti et al., 2022). Because the world stood at a standstill during the early stages of COVID-19, this could be an excellent chance for self-reflection and a revised version of environmental and ethical perceptions (Dumitras et al., 2021). Knowledge about food waste among many young people is crucial for developing an action plan. And which knowledge is acquired and how it is transferred are both critical. Nutrient courses, sustained courses, financial education, ethics, environmental conservation, and other curriculum content can be incorporated. There is a way to integrate not only held-to-account consumption but also financial knowledge courses. In this manner, the development of financial understanding is probably interconnected with knowledge development about sustainable products, and the two mindsets are mutually supportive in the end (Attiq et al., 2021). Reduction is a personal finance pillar that removes all waste, including wasted food. A savings plan's economic dosage requires innovative solutions to do more with less, including a food budget. Young people's opinions of the environmental effects of food waste can be enhanced in the same manner that transmission knowledge can. A theoretical basis must be combined with a practical learning approach. Young people must participate in thematic visits to environmentally impacted areas with limited access to water and local food (Vu et al., 2021). Their involvement in these harmed areas for a week or more will almost certainly lead to a shift in thoughts and a start embracing a pro-sustainability attitude. Volunteer assessments that necessitate young people to cook for disadvantaged groups on a tight budget and layout a menu with scarce funds will change their opinion on food waste.

13.6 CONCLUSION

This study shows food waste awareness in urban and rural areas after the COVID-19 pandemic. The finding suggests that the behavior of urban and rural people toward food waste is respectively positive, and they know how important it is to save food and utilize the leftover food. COVID-19 influences young minds toward online shopping of food items, which builds ethics regarding food waste. Elder ones are the one who also influences the young as well as the adults about food waste management and storage of food. Knowledge of the cause can be more controlled if government induces some policies about food waste management, which can be done through online platforms in urban and rural areas. Start giving importance to food from the start to kids in primary school, like in private and government schools. That's how we can contribute to raising awareness about household food waste, building attitudes, and shaping behaviors in urban and rural areas.

REFERENCES

Amicarelli, V., & Bux, C. (2021). Food waste in Italian households during the Covid-19 pandemic: A self-reporting approach. *Food Security*, *13*(1), 25–37. https://doi.org/10.1007/S12571-020-01121-Z/TABLES/5

Attiq, S., Chau, K. Y., Bashir, S., Habib, M. D., Azam, R. I., & Wong, W. K. (2021). Sustainability of household food waste reduction: A fresh insight on youth's emotional and cognitive behaviors. *International Journal of Environmental*

Research and Public Health, *18*(13), 7013. https://doi.org/10.3390/ IJERPH18137013

Baral, M. M., Mukherjee, S., Chittipaka, V., Srivastava, S. C., & Pal, S. K. (2022). Critical components for food wastage in food supply chain management. *Advances in Mechanical and Industrial Engineering*, 295–299. https://doi. org/10.1201/9781003216742-44

Baral, M. M., Mukherjee, S., Nagariya, R., Singh Patel, B., Pathak, A., & Chittipaka, V. (2022). Analysis of factors impacting firm performance of MSMEs: Lessons learnt from COVID-19. *Benchmarking: An International Journal*, ahead-of-print. https://doi.org/10.1108/BIJ-11-2021-0660

Beheshti, S., Heydari, J., & Sazvar, Z. (2022). Food waste recycling closed loop supply chain optimization through renting waste recycling facilities. *Sustainable Cities and Society*, *78*, 103644. https://doi.org/10.1016/J.SCS.2021.103644

Borg, K., Boulet, M., Karunasena, G., & Pearson, D. (2022). Segmenting households based on food waste behaviours and waste audit outcomes: Introducing over providers, under planners and considerate planners. *Journal of Cleaner Production*, *351*, 131589. https://doi.org/10.1016/J.JCLEPRO.2022.131589

Burlea-Schiopoiu, A., Ogarca, R. F., Barbu, C. M., Craciun, L., Baloi, I. C., & Mihai, L. S. (2021). The impact of COVID-19 pandemic on food waste behaviour of young people. *Journal of Cleaner Production*, *294*, 126333. https://doi. org/10.1016/J.JCLEPRO.2021.126333

Cakar, B. (2022). Bounce back of almost wasted food: Redistribution of fresh fruit and vegetables surpluses from Istanbul's supermarkets. *Journal of Cleaner Production*, *362*, 132325. https://doi.org/10.1016/J.JCLEPRO.2022.132325

Ch Hadjichambis, A., Al-Nuaimi, S. R., & Al-Ghamdi, S. G. (2022). Sustainable consumption and education for sustainability in higher education. *Sustainability*, *14*(12), 7255. https://doi.org/10.3390/SU14127255

Ciccullo, F., Fabbri, M., Abdelkafi, N., & Pero, M. (2022). Exploring the potential of business models for sustainability and big data for food waste reduction. *Journal of Cleaner Production*, *340*, 130673. https://doi.org/10.1016/J. JCLEPRO.2022.130673

Costa, F. H. de O., de Moraes, C. C., da Silva, A. L., Delai, I., Chaudhuri, A., & Pereira, C. R. (2022). Does resilience reduce food waste? Analysis of Brazilian supplier-retailer dyad. *Journal of Cleaner Production*, *338*, 130488. https://doi. org/10.1016/J.JCLEPRO.2022.130488

Cudjoe, D., Zhu, B., & Wang, H. (2022). Towards the realization of sustainable development goals: Benefits of hydrogen from biogas using food waste in China. *Journal of Cleaner Production*, *360*, 132161. https://doi.org/10.1016/J. JCLEPRO.2022.132161

Daliakopoulos, N., Chroni, C., Davison, N., Young, W., Ross, A., Cockerill, T., & Rajput, S. (2022). Investigating the impacts of behavioural-change interventions and COVID-19 on the food-waste-generation behaviours of catered students in the UK and India. *Sustainability*, *14*(9), 5486. https://doi.org/10.3390/SU14095486

Deng, Y., Chen, X., Adam, N. G. T. S., & Xu, J. (2022). A multi-objective optimization approach for clean treatment of food waste from an economic-environmental-social perspective: A case study from China. *Journal of Cleaner Production*, *357*, 131559. https://doi.org/10.1016/J.JCLEPRO.2022.131559

Diekmann, L., Germelmann, C. C., Jahn, S., & Furchheim, P. (2021). Leftover consumption as a means of food waste reduction in public space? Qualitative insights from online discussions. *Sustainability*, *13*(24), 13564. https://doi.org/10.3390/SU132413564

Dumitras, D. E., Harun, R., Arion, F. H., Chiciudean, D. I., Kovacs, E., Oroian, C. F., Porutiu, A., & Muresan, I. C. (2021). Food consumption patterns in Romania during the COVID-19 pandemic. *Foods*, *10*(11), 2712. https://doi.org/10.3390/FOODS10112712

Durán-Sandoval, D., Durán-Romero, G., & López, A. M. (2021). Achieving the food security strategy by quantifying food loss and waste: A case study of the Chinese economy. *Sustainability*, *13*(21), 12259. https://doi.org/10.3390/SU132112259

Fan, L., Ellison, B., & Wilson, N. L. W. (2022). What Food waste solutions do people support? *Journal of Cleaner Production*, *330*, 129907. https://doi.org/10.1016/J.JCLEPRO.2021.129907

Filimonau, V., & Uddin, R. (2021). Food waste management in chain-affiliated and independent consumers' places: A preliminary and exploratory study. *Journal of Cleaner Production*, *319*, 128721. https://doi.org/10.1016/J.JCLEPRO.2021.128721

Filimonau, V., Vi, L. H., Beer, S., & Ermolaev, V. A. (2022). The Covid-19 pandemic and food consumption at home and away: An exploratory study of English households. *Socio-Economic Planning Sciences*, *82*, 101125. https://doi.org/10.1016/J.SEPS.2021.101125

Iranmanesh, M., Ghobakhloo, M., Nilashi, M., Tseng, M.-L., Senali, M. G., & Abbasi, G. A. (2022). Impacts of the COVID-19 pandemic on household food waste behaviour: A systematic review. *Appetite*, *176*, 106127. https://doi.org/10.1016/J.APPET.2022.106127

Jenkins, E. L., Brennan, L., Molenaar, A., & McCaffrey, T. A. (2022). Exploring the application of social media in food waste campaigns and interventions: A systematic scoping review of the academic and grey literature. *Journal of Cleaner Production*, *360*, 132068. https://doi.org/10.1016/J.JCLEPRO.2022.132068

Kharola, S., Ram, M., Kumar Mangla, S., Goyal, N., Nautiyal, O. P., Pant, D., & Kazancoglu, Y. (2022). Exploring the green waste management problem in food supply chains: A circular economy context. *Journal of Cleaner Production*, *351*, 131355. https://doi.org/10.1016/J.JCLEPRO.2022.131355

Lu, L. C., Chiu, S. Y., Chiu, Y. Ho, & Chang, T. H. (2022). Three-stage circular efficiency evaluation of agricultural food production, food consumption, and food waste recycling in EU countries. *Journal of Cleaner Production*, *343*, 130870. https://doi.org/10.1016/J.JCLEPRO.2022.130870

Massari, S., Principato, L., Antonelli, M., & Pratesi, C. A. (2022). Learning from and designing after pandemics. CEASE: A design thinking approach to maintaining food consumer behaviour and achieving zero waste. *Socio-Economic Planning Sciences*, *82*, 101143. https://doi.org/10.1016/J.SEPS.2021.101143

Mukherjee, S., Baral, M. M., Chittipaka, V., & Pal, S. K. (2022). Addressing the strategies for the sustainable supply chain in post-COVID-19 pandemic. In *Making complex decisions toward revamping supply chains amid COVID-19 outbreak* (pp. 69–86). CRC Press. https://doi.org/10.1201/9781003150084-4

Mukherjee, S., Baral, M. M., Pal, S. K., Chittipaka, V., Roy, R., & Alam, K. (2022). Humanoid robot in healthcare: A Systematic Review and Future Research Directions. *2022 International Conference on Machine Learning, Big Data, Cloud and Parallel Computing (COM-IT-CON)*, 822–826. https://doi.org/10.1109/COM-IT-CON54601.2022.9850577

Mukherjee, S., & Chittipaka, V. (2021). Analysing the adoption of intelligent agent technology in food supply chain management: An empirical evidence. *FIIB Business Review*. https://doi.org/10.1177/23197145211059243

Mukherjee, S., Chittipaka, V., & Baral, M. M. (2022). Addressing and modeling the challenges faced in the implementation of blockchain technology in the food and agriculture supply chain. *Blockchain Technologies and Applications for Digital Governance*, 151–179. https://doi.org/10.4018/978-1-7998-8493-4.ch007

Mukherjee, S., Chittipaka, V., Baral, M. M., Pal, S. K., & Rana, S. (2022). Impact of artificial intelligence in the healthcare sector. *Artificial Intelligence and Industry 4.0*, 23–54. https://doi.org/10.1016/B978-0-323-88468-6.00001-2

Mukherjee, S., Chittipaka, V., Baral, M. M., & Srivastava, S. C. (2022). Can the supply chain of Indian SMEs adopt the technologies of industry 4.0? *Advances in Mechanical and Industrial Engineering*, 300–304. https://doi.org/10.1201/9781003216742-45

Mukherjee, S., Venkataiah, C., Baral, M. M., & Pal, S. K. (2022). Analyzing the factors that will impact the supply chain of the COVID-19 vaccine: A structural equation modeling approach. *Journal of Statistics and Management Systems*, 1–16. https://doi.org/10.1080/09720510.2021.1966955

Münch, C., von der Gracht, H. A., & Hartmann, E. (2021). The future role of reverse logistics as a tool for sustainability in food supply chains: a Delphi-based scenario study. *Supply Chain Management*, ahead-of-print. https://doi.org/10.1108/SCM-06-2021-0291/FULL/PDF

Obuobi, B., Zhang, Y., Adu-Gyamfi, G., Nketiah, E., Grant, M. K., Adjei, M., & Cudjoe, D. (2022). Fruits and vegetable waste management behavior among retailers in Kumasi, Ghana. *Journal of Retailing and Consumer Services*, 67, 102971. https://doi.org/10.1016/J.JRETCONSER.2022.102971

Roy, R., Baral, M. M., Pal, S. K., Kumar, S., Mukherjee, S., & Jana, B. (2022). Discussing the present, past, and future of Machine learning techniques in livestock farming: A systematic literature review. *2022 International Conference on Machine Learning, Big Data, Cloud and Parallel Computing (COM-IT-CON)*, 179–183. https://doi.org/10.1109/COM-IT-CON54601.2022.9850749

Strotmann, C., Baur, V., Börnert, N., & Gerwin, P. (2022). Generation and prevention of food waste in the German food service sector in the COVID-19 pandemic—digital approaches to encounter the pandemic related crisis. *Socio-Economic Planning Sciences*, 82, 101104. https://doi.org/10.1016/J.SEPS.2021.101104

Vasseur, L., Vanvolkenburg, H., Vandeplas, I., Touré, K., Sanfo, S., & Baldé, F. L. (2021). The effects of pandemics on the vulnerability of food security in West Africa—a scoping review. *Sustainability*, 13(22), 12888. https://doi.org/10.3390/SU132212888

Vu, H. L., Ng, K. T. W., Richter, A., Karimi, N., & Kabir, G. (2021). Modeling of municipal waste disposal rates during COVID-19 using separated waste fraction models. *Science of the Total Environment*, 789, 148024. https://doi.org/10.1016/J.SCITOTENV.2021.148024

Yan, J., Luo, F., Wu, lingyao, Ou, Y., Gong, C., Hao, T., Huang, L., Chen, Y., Long, J., Xiao, T., & Zhang, H. (2022). Cost-effective desulfurization of acid mine drainage with food waste as an external carbon source: A pilot-scale and long-term study. *Journal of Cleaner Production*, 361, 132174. https://doi.org/10.1016/J.JCLEPRO.2022.132174

Yi-Chi Chang, Y., Lin, J.-H., & Hsiao, C.-H. (2022). Examining effective means to reduce food waste behaviour in buffet restaurants. *International Journal of Gastronomy and Food Science*, 29, 100554. https://doi.org/10.1016/J.IJGFS.2022.100554

Yousefi, M., Oskoei, V., Jonidi Jafari, A., Farzadkia, M., Hasham Firooz, M., Abdollahinejad, B., & Torkashvand, J. (2021). Municipal solid waste management during COVID-19 pandemic: effects and repercussions. *Environmental Science and Pollution Research*, 28(25), 32200–32209. https://doi.org/10.1007/S11356-021-14214-9/TABLES/2

Chapter 14

A conceptual study of association factors contributing to the stress level of parents during the pandemic era in NCR

Nitendra Kumar, Pooja Singh, Santosh Kumar, Surya Kant Pal, Khursheed Alam, and Mahesh Kumar Jayaswal

14.1 INTRODUCTION

Considering the current situation of COVID-19, the whole economy is suffering due to every sector, which has badly affected the country's education system. Though online classes are running currently, there are various challenges like modernization, support, or coordination between students, teachers, and parents for a successful e-learning process [1]. There has been a radical change in traditional education concepts within the last couple of years. All thanks to the new technology and rising internet facilities, sitting in the classroom and learning is not the only option left. Now we can sit and learn virtually anywhere and at any time. Online education is a savior for children within the age group of 3 to 21. This epidemic teaches parents about added responsibilities such as extra care toward their child's growth, mental awareness, and knowledge level. In nuclear families, where people rely on house help and daycare facilities, it is quite challenging for them to manage all the responsibilities alone and the education because most working families rely on tuition and extracurricular activities such as swimming and karate classes for their engagements. Online education is another big milestone in their pandemic life period. Although teachers are trying to provide their assistance with new methodologies for children learning, the parent's part is more challenging as they are the ones who make them understand better and unified concepts.

14.2 REVIEW OF LITERATURE

The implementation part of online education is considered a challenging part for the children and parents. To promote learning skills, parents are trying to take help from Google and online websites for their indulgence [2]. During the pandemic, multiple factors are affecting the stress level of

DOI: 10.1201/9781003462422-14

parents. The factors are good marriage, cordial relations with parents, children's education, and many more [3].

Negligence and compromised parenting lead to children in stress; this is not uncommon in our present-day society. Child nurturing has been viewed as an essential component of parenting in nuclear families. However, the ongoing pandemic is causing new stress for parents. There are numerous causes of stress in the present condition, but health and economic conditions are prioritized [4]. The conditions of life have changed simultaneously. The importance and role of parents in education have now been on priority and crucial. The concepts of homework and online classes have paved the parents toward more attentive and supportive in their children's assignments. Now the methodology and concern also changed throughout its entirety, especially for toddlers and preschoolers [5].

Before the pandemic, the culture of online education was growing steadily and simultaneously but now as the digital world expanded the adoption of the internet and technology developed spontaneously. There is an increasing demand for digitally skilled persons along with global communication in the mainstream. Virtual reality has shifted our attention from personal meetings to video communication. Our society is bound to adopt the new challenges imposed by COVID-19, such as embracing the new culture of technology [6,7]. From business meetings to education, all the schedules are classified as virtual reality, the education and meetings are inbound to be performed on Google Meet and Zoom, but healthcare and other essentials of households are possible during this outbreak [8]. Organizations are using telecommunications methods to adopt the new culture of the pandemic. On the other hand, telemarketing is gaining popularity to grab customers' attention. In the education sector, where the gathering of students is not essential, there is a compulsion for online learning.

Stress is all the way related to the physical growth of children. Mental stress can be negative or positive, sometimes embracing a new culture and paving toward a new beginning also causes stress, but that is surely positive, while negative stress leads to health deterioration. Parents play a pivotal role in managing the stress of their children. Adequate stress for managing things and completing tasks on time can be good, but over time, excessive stress can affect the physical growth of children and their parents [9, 10]. A child gradually learns how to respond to and manage stress levels. There are some conditions where the parents manage to tackle the stress while a child cannot, and there are vice versa conditions also. There are a few conditions for children's stress, such as pain, illness, school grades, sports competition, friends problems, bullying, homesickness, negative thoughts about self-incompetence, body changes, and family problems [6, 9, 10, 11].

Many stress conditions are temporary and hence can be resolved by the children themselves. Still, a few conditions can prevail in children's memory and worsen their physical growth. The symptoms include reduced appetite, stomach ache, nightmares, illness, etc. In this case, the parents need to understand the root cause of the problem and simultaneously need to resolve the issue through proper counseling of children [12]. Mental health and stability play a pivotal role in children's upbringing. The parents deal with the stress; it is obvious to the children that their parents are facing some issues. The changed behavior of parents suddenly gets noticed by the children. Parents can manage their issues by adopting various stress-buster programs such as exercising, Yoga, Music, and other activities. Parental stress pulls the attention of their children and hindrance their upbringing [11, 12, 13, 14].

14.3 MATERIAL AND METHODS

From the extensive literature review, it has been identified that three factors are important in measuring parents' stress levels toward their child's online education. The questionnaire is based on the standardized tool [15, 16]. These two standard tools are taken under the research study. The Likert scale method is used to analyze the questionnaire. The chi-square method is used to obtain the association between various reasons for concerns and their impact on parents' stress levels. The questions were asked to the parents about their perceptions of experiences and their valid concern about the epidemic. For each point of concern, parents were asked to rate their answers on the Likert Scale of 1–5 [8].

The potential participants found with the variables and indulgence in online education, physical growth, and mental peace during the pandemic [17]. The analysis is done through SPSS 22.0. The judgmental sampling method was used to fill the questionnaire; approximately 300 questionnaires were floated in WhatsApp groups (formal and informal), 100 have not filled, 20 were found obsolete, and 30 were vague and hard to conclude; therefore, 150 questionnaires were found useful, and on that basis, reliability and validity were tested. The Cronbach alpha was coming to be 0.75 and hence considered acceptable for the study. The different hypotheses are measured through different techniques.

Hypothesis:

H_{01}: Online education is not an association variable of parents' stress level during COVID-19 in NCR.

H_{02}: Mental peace is not an association variable of parents' stress level during COVID-19 in NCR.

H_{03}: Physical growth is not an association variable of parents' stress level during COVID-19 in NCR.

14.4 DATA ANALYSIS

We have used some tools to build the questionnaire to analyze the variable of the study. Initially, there were 30 questions, ten were incomplete, and eight were obsolete and vague. Therefore, we deleted those questions. Out of that, 12 questions were asked of the respondents.

Steps for developing the Likert Scale:

- Define what is your center of research.
- Initiation of items based on the Likert scale.
- Rate the items on a Likert scale.
- Manage the scale.

While performing an analysis using Likert analysis, it is convenient to modify the questions in the initial stage of the analysis, rather than collecting data randomly and then deciding on a measure of analysis. The analysis of data is based on the format of the questionnaire. Subsequently, reliability and validity tests can be applied to analyze the data obtained after collecting the data in the form of a questionnaire using the Likert scales such as the Chi-square test, symmetric measures, and KMO, and Bartlett's test.

A chi-square formula is used to drag the relationship between various study variables.

$$\chi^2 = \Sigma \frac{\left(O_i - E_i\right)^2}{E_i}$$

where O_i is the observed value and E_i is the expected value.

This test is used to explore the relationship between two categorical variables. Each of these variables can have two or more categories. It is based on a cross-tabulation table, with cases classified according to the categories in each variable.

- **Click on** Analyze\Descriptive Statistics\Crosstabs
- Move one of your categorical variables into the marked box Row(s).
- Move the other categorical variable into the box marked Column(s).
- Click on the Statistics button and tick Chi-square and Phi and Cramer's V. Click on Continue.
- Click on the Cells button: in the Counts box, make sure there is a tick for Observed; in the Percentage section, click on the Row, Column, and Total boxes.
- Click on Continue and OK.

H_{01}: Online education is not an association variable of parents' stress level during COVID-19 in NCR.

Table 14.1 Chi-Square Tests

	Value	Df	Asymp. sig. (two-sided)
Pearson Chi-Square	35.444[a]	16	.003
Likelihood Ratio	30.020	16	.018
Linear-by-Linear Association	12.925	1	.000
N of Valid Cases	130		

Table 14.2 Symmetric Measures

		Value	Asymp. std. error[a]	Approx. T[b]	Approx. sig.
Interval by Interval	Pearson's R	.317	.090	3.775	.000[c]
Ordinal by Ordinal	Spearman Correlation	.296	.083	3.511	.001[c]
N of Valid Cases		130			

Online education is the biggest stress maker for parents whose children lie in the age group of 3–12. The likelihood value comes to be 0.018, which is less than 0.05, proving that online education is prevalent with the level of understanding, level of intelligence, commitment, and experience must remain present whenever the parents' stress level is calculated. The linear-by-linear association test reveals that all four are the sub-variables of stress measurement.

H_{02}: Mental peace is not an association variable of parents' stress level during COVID-19 in NCR.

Yoga and meditation are found to be associated with peace, and quality time is directly linked to mental satisfaction toward work and the child. The value of symmetric measures, which is stated through the Pearson and Spearman Correlation, is 0.00 and 0.001, revealing that their mental peace impacts the measurement of parents' stress levels during online learning.

H_{03}: Physical growth is not an association variable of parents' stress level during COVID-19 in NCR.

Kaiser–Meyer–Olkin (KMO) and Bartlett's tests **measure the strength of the relationship among variables**. The KMO measures the sampling adequacy, which should be greater than 0.5 for satisfactory factor analysis to proceed.

In SPSS: Run factor analysis (analyze > dimension reduction > factor) and check the box for "KMO and Bartlett's test of sphericity".

The formula for the KMO test is as follows:

$$MO_j = \frac{\sum_{i \neq j} r_{ij}^2}{\sum_{i \neq j} r_{ij}^2 + \sum_{i \neq j} u}$$

Table 14.3 KMO and Bartlett's Tests

Kaiser–Meyer–Olkin Measure of Sampling Adequacy.		.677
Bartlett's Test of Sphericity	Approx. Chi-Square	378.027
	Df	78
	Sig.	.000

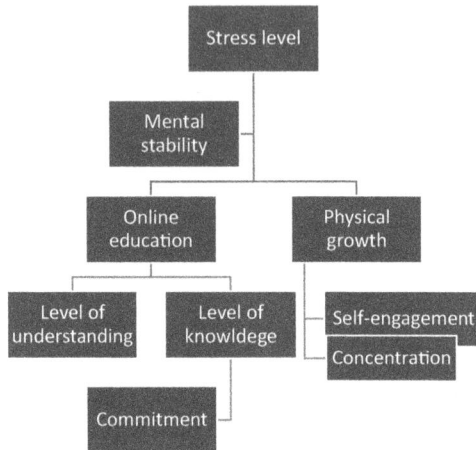

Figure 14.1 Conceptual model of factors affecting stress level of parents during the pandemic.

where $R = [r_{ij}]$ is the correlation matrix and $U = [u_{ij}]$ is the partial covariance matrix.

The impact of physical growth is reflected through the Kaiser–Meyer method to know the adequacy of the correlation method. The significance values of KMO reflect that Physical growth is an association variable of stress level. Here the research pointed out the physical growth of the children during COVID-19, which reflects that growth is dependent on self-engagement and concentration. Therefore, stress level is associated with self-engagement and a child's concentration on education.

14.5 RESULTS AND DISCUSSIONS

Our findings highlighted that, during the lockdown, parents suffered from a much higher rate of stress, anxiety, and depression than at the normal time [9]. Level of understanding, level of knowledge, and commitment toward education are highly associated sub-variables of online education. Online education is raising the understanding level, and at the same time, it is a compulsion for the parents to sit along with their wards, and it is not providing the same exposure as offline studies. Parents are not trying to multitask and are more responsible.

Parents were also facing the trouble of keeping their wards inside to prevent them from COVID-19; physical growth has shown a strong correlation with self-engagement and concentration [10, 12, 17, 18]. Mental peace is necessary for reducing frustration through yoga, exercises, and other entertainment engagements, but mental peace has not shown any direct associations with stress level measurements. The continuous pandemic, which has no endpoint, is giving frustration to the entire society, but parents suffer the most from the current situation. The government is strictly posing the isolation measures to follow, but the public is so desperate to travel the world. The common public does just not understand the ill effect of the situation. The parents are struggling on various grounds, from online teaching to keeping pace with their children's growth everything is just under the mental pressure of the parents. In the growth period of children, where they have to make new friends, exercise, and play on the grounds, kids are suffering and bound to play indoor games.

REFERENCES

[1] R. Watermeyer, T. Crick, C. Knight, and J. Goodall, "COVID-19 and digital disruption in UK universities: Afflictions and affordances of emergency online migration", *Higher Education* 81(3), 623–641, 2021.

[2] E. Susilowati, and M. Azzasyofia, "The parents stress level in facing children study from home in the early of Covid-19 pandemic in Indonesia", *International Journal of Science and Society* 2(3), 1–12, 2020.

[3] M. Wu, W. Xu, Y. Yao, L. Zhang, L. Guo, J. Fan, and J. Chen, "Mental health status of students' parents during COVID-19 pandemic and its influence factors", *General Psychiatry* 33(4), 2020.

[4] S. M. Brown, J. R. Doom, S. Lechuga-Peña, S. E. Watamura, and T. Koppels, "Stress and parenting during the global COVID-19 pandemic", *Child Abuse & Neglect*, 110(10), 46–99, 2020.

[5] G. Wang, Y. Zhang, J. Zhao, J. Zhang, and F., Jiang, "Mitigate the effects of home confinement on children during the COVID-19 outbreak", *Lancet* 395, 945–947, 2020.

[6] G. Madaan, H.R. Swapna, A. Kumar, A. Singh, and A. David, "Enactment of sustainable technovations on healthcare sectors", *Asia Pacific Journal of Health Management* 16(3), 184–192, 2021.

[7] A. Kumar, G. Madaan, P. Sharma, and A. Kumar, "Application of disruptive technologies on environmental health: An overview of artificial intelligence, blockchain and internet of things", *Asia Pacific Journal of Health Management* 16(4), 251–259, 2021.

[8] S.K. Pal, S. Mukherjee, M.M. Baral, and S. Aggarwal, "Problems of big data Adoption in the healthcare industries", *Asia Pacific Journal of Health Management* 16(4), 282–287, 2021.

[9] M. Mosanya, "Buffering academic stress during the COVID-19 pandemic related social isolation: Grit and growth mindset as protective factors

against the impact of loneliness", *International Journal of Applied Positive Psychology*, 6(2), 159–174, 2021.

[10] M. Spinelli, F. Lionetti, M. Pastore, and M. Fasolo, "Parents' stress and children's psychological problems in families facing the COVID-19 outbreak in Italy", *Frontiers in Psychology*, 11, 1713, 2020.

[11] P. Pujari, P. Pujari, and A. Kumar, "Impact of covid-19 on the mental health of healthcare workers: Predisposing factors, prevalence and supportive strategies", *Asia Pacific Journal of Health Management* 16(4), 260–265, 2021.

[12] R. Montirosso, E. Mascheroni, E. Guida, C. Piazza, M.E. Sali, M. Molteni, and G. Reni, "Stress symptoms and resilience factors in children with neurodevelopmental disabilities and their parents during the COVID-19 pandemic", *Health Psychology*, 40(7), 428, 2021.

[13] R. Roy, M.M. Baral, S.K. Pal, S. Kumar, S. Mukherjee, and B. Jana, "Discussing the present, past, and future of Machine learning techniques in livestock farming: A systematic literature review", *2022 International Conference on Machine Learning, Big Data, Cloud and Parallel Computing (COM-IT-CON)*, 179–183, 2022. doi:10.1109/COM-IT-CON54601.2022.9850749.

[14] S. Mukherjee, M.M. Baral, S.K. Pal, V. Chittipaka, R. Roy, and K. Alam, "Humanoid robot in healthcare: A systematic review and future research directions", *2022 International Conference on Machine Learning, Big Data, Cloud and Parallel Computing (COM-IT-CON)*, 822–826, 2022. doi:10.1109/COM-IT-CON54601.2022.9850577.

[15] J. Akbari, R. Akbari, M. Shakerian, and B. Mahaki, "Job demand-control and job stress at work: A cross-sectional study among prison staff", *Journal of Education and Health Promotion*, 6, 2017.

[16] M.L. Carey, A.C. Zucca, M.A.G. Freund, J. Bryant, A. Herrmann, and B.J. Roberts, *Systematic review of barriers and enablers to the delivery of palliative care by primary care practitioners*, Published in Palliative Medicine [In Press], 2019.

[17] S.K. Pal, and A.K. Pal, "The impact of increase in COVID-19 cases with exceptional situation to SDG: Good health and well-being", *Journal of Statistics & Management Systems* 24(1), 209–228, 2021.

[18] E. Susilowati, and M. Azzasyofia, "The parents stress level in facing children study from home in the early of Covid-19 pandemic in Indonesia", *International Journal of Science and Society* 2(3), 1–12, 2020.

Index

For Product Safety Concerns and Information please contact our EU
representative GPSR@taylorandfrancis.com
Taylor & Francis Verlag GmbH, Kaufingerstraße 24, 80331 München, Germany

www.ingramcontent.com/pod-product-compliance
Lightning Source LLC
Chambersburg PA
CBHW060352220326
41598CB00023B/2891